PENNSYLVANIA COLLEGE OF TECHNOLOGY LIBRARY

5 0608 01182965 1

D0992627

INSTRUCTI.
CHECKOUT TO CURRENT FACULTY,
STAFF, AND STUDENTS ONLY

2015 IMC®

INTERNATIONAL
Mechanical Code®

CODE ALERT!

Sign up now to receive critical code updates and free access to videos,
book excerpts and training resources.

Signup is easy, subscribe now! www.iccsafe.org/alerts

Madigan Library
Pennsylvania College
of Technology

One College Avenue
Williamsport, PA 17701-5799

NOV 0 1 2019

2015 International Mechanical Code®

Date of First Publication: May 30, 2014

First Printing: May 2014
Second Printing: September 2015
Third Printing: November 2015

ISBN: 978-1-60983-479-1 (soft-cover edition)
ISBN: 978-1-60983-478-4 (loose-leaf edition)

COPYRIGHT © 2014
by
INTERNATIONAL CODE COUNCIL, INC.

ALL RIGHTS RESERVED. This 2015 *International Mechanical Code*® is a copyrighted work owned by the International Code Council, Inc. ("ICC"). Without advance written permission from the ICC, no part of this book may be reproduced, distributed or transmitted in any form or by any means, including, without limitation, electronic, optical or mechanical means (by way of example, and not limitation, photocopying, or recording by or in an information storage retrieval system). For information on use rights and permissions, please contact: ICC Publications, 4051 Flossmoor Road, Country Club Hills, IL 60478. Phone 1-888-ICC-SAFE (422-7233).

Trademarks: "International Code Council," the "International Code Council" logo, "ICC," the "ICC" logo, "International Mechanical Code," "IMC" and other names and trademarks appearing in this book are registered trademarks of the International Code Council, Inc., and/or its licensors (as applicable), and may not be used without permission.

T023300
PRINTED IN THE USA

PREFACE

Introduction

Internationally, code officials recognize the need for a modern, up-to-date mechanical code addressing the design and installation of mechanical systems through requirements emphasizing performance. The *International Mechanical Code*®, in this 2015 edition, is designed to meet these needs through model code regulations that safeguard the public health and safety in all communities, large and small.

This comprehensive mechanical code establishes minimum regulations for mechanical systems using prescriptive and performance-related provisions. It is founded on broad-based principles that make possible the use of new materials and new mechanical designs. This 2015 edition is fully compatible with all of the *International Codes* (I-Codes®) published by the International Code Council (ICC), including the *International Building Code*®, *International Energy Conservation Code*®, *International Existing Building Code*®, *International Fire Code*®, *International Fuel Gas Code*®, *International Green Construction Code*®, *International Plumbing Code*®, ICC *Performance Code*®, *International Private Sewage Disposal Code*®, *International Property Maintenance Code*®, *International Residential Code*®, *International Swimming Pool and Spa Code*™, *International Urban-Wildland Interface Code*® and *International Zoning Code*®.

The *International Mechanical Code* provisions provide many benefits, among which is the model code development process that offers an international forum for mechanical professionals to discuss performance and prescriptive code requirements. This forum provides an excellent arena to debate proposed revisions. This model code also encourages international consistency in the application of provisions.

Development

The first edition of the *International Mechanical Code* (1996) was the culmination of an effort initiated in 1994 by a development committee appointed by the ICC and consisting of representatives of the three statutory members of the International Code Council at that time, including: Building Officials and Code Administrators International, Inc. (BOCA), International Conference of Building Officials (ICBO) and Southern Building Code Congress International (SBCCI). The intent was to draft a comprehensive set of regulations for mechanical systems consistent with and inclusive of the scope of the existing model codes. Technical content of the latest model codes promulgated by BOCA, ICBO and SBCCI was utilized as the basis for the development. This 2015 edition presents the code as originally issued, with changes approved through the ICC Code Development Process through 2013. A new edition such as this is promulgated every 3 years.

This code is founded on principles intended to establish provisions consistent with the scope of a mechanical code that adequately protects public health, safety and welfare; provisions that do not unnecessarily increase construction costs; provisions that do not restrict the use of new materials, products or methods of construction; and provisions that do not give preferential treatment to particular types or classes of materials, products or methods of construction.

Adoption

The International Code Council maintains a copyright in all of its codes and standards. Maintaining copyright allows ICC to fund its mission through sales of books, in both print and electronic formats. The *International Mechanical Code* is designed for adoption and use by jurisdictions that recognize and acknowledge the ICC's copyright in the code, and further acknowledge the substantial shared value of the public/private partnership for code development between jurisdictions and the ICC.

The ICC also recognizes the need for jurisdictions to make laws available to the public. All ICC codes and ICC standards, along with the laws of many jurisdictions, are available for free in a non-downloadable form on the ICC's website. Jurisdictions should contact the ICC at adoptions@icc-safe.org to learn how to adopt and distribute laws based on the *International Mechanical Code* in a manner that provides necessary access, while maintaining the ICC's copyright.

Maintenance

The *International Mechanical Code* is kept up to date through the review of proposed changes submitted by code enforcing officials, industry representatives, design professionals and other interested parties. Proposed changes are carefully considered through an open code development process in which all interested and affected parties may participate.

The contents of this work are subject to change through both the code development cycles and the governmental body that enacts the code into law. For more information regarding the code development process, contact the Codes and Standards Development Department of the International Code Council.

While the development procedure of the *International Mechanical Code* ensures the highest degree of care, the ICC, its members and those participating in the development of this code do not accept any liability resulting from compliance or noncompliance with the provisions because the ICC does not have the power or authority to police or enforce compliance with the contents of this code. Only the governmental body that enacts the code into law has such authority.

Code Development Committee Responsibilities
(Letter Designations in Front of Section Numbers)

In each code development cycle, proposed changes to this code are considered at the Committee Action Hearing by the International Mechanical Code Development Committee, whose action constitutes a recommendation to the voting membership for final action on the proposed change. Proposed changes to a code section that has a number beginning with a letter in brackets are considered by a different code development committee. For example, proposed changes to code sections that have [BG] in front of them (e.g., [BG] 309.1) are considered by the IBC — General Code Development Committee at the Committee Action Hearing.

The content of sections in this code that begin with a letter designation is maintained by another code development committee in accordance with the following:

[A] = Administrative Code Development Committee;

[BF] = IBC — Fire Safety Code Development Committee;

[BS] = IBC — Structural Code Development Committee;

[BG] = IBC — General Code Development Committee;

[E] = International Energy Conservation Code Development Committee;

[F] = International Fire Code Development Committee; and

[FG] = International Fuel Gas Code Development Committee.

For the development of the 2018 edition of the I-Codes, there will be three groups of code development committees and they will meet in separate years. Note that these are tentative groupings.

Group A Codes (Heard in 2015, Code Change Proposals Deadline: January 12, 2015)	Group B Codes (Heard in 2016, Code Change Proposals Deadline: January 11, 2016)	Group C Codes (Heard in 2017, Code Change Proposals Deadline: January 11, 2017)

International Building Code
  – Fire Safety (Chapters 7, 8, 9, 14, 26)
  – Means of Egress (Chapters 10, 11, Appendix E)
  – General (Chapters 2-6, 12, 27-33, Appendices A, B, C, D, K)
	Administrative Provisions (Chapter 1 of all codes except IRC and IECC, administrative updates to currently referenced standards, and designated definitions)	International Green Construction Code

International Fuel Gas Code	International Building Code – Structural (Chapters 15-25, Appendices F, G, H, I, J, L, M)	
International Existing Building Code	International Energy Conservation Code	
International Mechanical Code	International Fire Code	
International Plumbing Code	International Residential Code – IRC-B (Chapters 1-10, Appendices E, F, H, J, K, L, M, O, R, S, T, U)	
International Private Sewage Disposal Code	International Wildland-Urban Interface Code	
International Property Maintenance Code		

International Residential Code
  – IRC-Mechanical (Chapters 12-24)
  – IRC-Plumbing (Chapters 25-33, Appendices G, I, N, P)		

International Swimming Pool and Spa Code		
International Zoning Code		

Note: Proposed changes to the ICC *Performance Code* will be heard by the code development committee noted in brackets [ ] in the text of the code.

Code change proposals submitted for code sections that have a letter designation in front of them will be heard by the respective committee responsible for such code sections. Because different committees hold code development hearings in different years, proposals for this code will be heard by committees in both the 2015 (Group A) and the 2016 (Group B) code development cycles.

For example, every section of Chapter 1 of this code is designated as the responsibility of the Administrative Code Development Committee, and that committee is part of the Group B code hearings. This committee will conduct its code development hearings in 2016 to consider all code change proposals for Chapter 1 of this code and proposals for Chapter 1 of all I-Codes except the *International Energy Conservation Code*, the ICC *Performance Code* and the *International Residential Code*. Therefore, any proposals received for Chapter 1 of this code will be deferred for consideration in 2016 by the Administrative Code Development Committee.

Another example is Section 606.4 of this code which is designated as the responsibility of the International Fire Code Development Committee. This committee will conduct its code development hearings in 2016 to consider code change proposals in its purview, which includes any proposals to Section 606.4.

In some cases, another committee in Group A will be responsible for a section of this code. For example, Section 607 has a [BF] in front of the numbered sections, indicating that these sections of the code are the responsibility of one of the International Building Code Development Committees. The International Building Code is in Group A; therefore, any code change proposals to this section will be due before the Group A deadline of January 2015, and these code change proposals will be assigned to the appropriate International Building Code Development Committee for consideration.

It is very important that anyone submitting code change proposals understand which code development committee is responsible for the section of the code that is the subject of the code change proposal. For further information on the code development committee responsibilities, please visit the ICC website at www.iccsafe.org/scoping.

Marginal Markings

Solid vertical lines in the margins within the body of the code indicate a technical change from the requirements of the 2012 edition. Deletion indicators in the form of an arrow (➡) are provided in the margin where an entire section, paragraph, exception or table has been deleted or an item in a list of items or a table has been deleted.

A single asterisk [*] placed in the margin indicates that text or a table has been relocated within the code. A double asterisk [**] placed in the margin indicates that the text or table immediately following it has been relocated there from elsewhere in the code. The following table indicates such relocations in the 2015 edition of the *International Mechanical Code*.

2015 LOCATION	2012 LOCATION
None	None

Italicized Terms

Selected terms set forth in Chapter 2, Definitions, are italicized where they appear in code text. Such terms are not italicized where the definition set forth in Chapter 2 does not impart the intended meaning in the use of the term. The terms selected have definitions that the user should read carefully to facilitate better understanding of the code.

EFFECTIVE USE OF THE INTERNATIONAL MECHANICAL CODE

The *International Mechanical Code*® (IMC®) is a model code that regulates the design and installation of mechanical systems, appliances, appliance venting, duct and ventilation systems, combustion air provisions, hydronic systems and solar systems. The purpose of the code is to establish the minimum acceptable level of safety and to protect life and property from the potential dangers associated with the installation and operation of mechanical systems. The code also protects the personnel that install, maintain, service and replace the systems and appliances addressed by this code.

The IMC is primarily a prescriptive code with some performance text. The code relies heavily on product specifications and listings to provide much of the appliance and equipment installation requirements. The general Section 105.2 and the exception to Section 403.2 allow designs and installations to be performed by approved engineering methods as alternatives to the prescriptive methods in the code.

The format of the IMC allows each chapter to be devoted to a particular subject with the exception of Chapter 3, which contains general subject matters that are not extensive enough to warrant their own independent chapter.

Chapter 1 Scope and Administration. Chapter 1 establishes the limits of applicability of the code and describes how the code is to be applied and enforced. A mechanical code, like any other code, is intended to be adopted as a legally enforceable document and it cannot be effective without adequate provisions for its administration and enforcement. The provisions of Chapter 1 establish the authority and duties of the code official appointed by the jurisdiction having authority and also establish the rights and privileges of the design professional, contractor and property owner.

Chapter 2 Definitions. Chapter 2 is the repository of the definitions of terms used in the body of the code. Codes are technical documents and every word and term can impact the meaning of the code text and the intended results. The code often uses terms that have a unique meaning in the code and the code meaning can differ substantially from the ordinarily understood meaning of the term as used outside of the code.

The terms defined in Chapter 2 are deemed to be of prime importance in establishing the meaning and intent of the code text that uses the terms. The user of the code should be familiar with and consult this chapter because the definitions are essential to the correct interpretation of the code and because the user may not be aware that a term is defined.

Chapter 3 General Regulations. Chapter 3 contains broadly applicable requirements related to appliance location and installation, appliance and systems access, protection of structural elements, condensate disposal and clearances to combustibles, among others.

Chapter 4 Ventilation. Chapter 4 includes means for protecting building occupant health by controlling the quality of indoor air and protecting property from the effects of inadequate ventilation. In some cases, ventilation is required to prevent or reduce a health hazard by removing contaminants at their source.

Ventilation is both necessary and desirable for the control of air contaminants, moisture and temperature. Habitable and occupiable spaces are ventilated to promote a healthy and comfortable environment for the occupants. Uninhabited and unoccupied spaces are ventilated to protect the building structure from the harmful effects of excessive humidity and heat. Ventilation of specific occupancies is necessary to minimize the potential for toxic or otherwise harmful substances to reach dangerously high concentrations in air.

Chapter 5 Exhaust Systems. Chapter 5 provides guidelines for reasonable protection of life, property and health from the hazards associated with exhaust systems, air contaminants and smoke development in the event of a fire. In most cases, these hazards involve materials and gases that are flammable, explosive, toxic or otherwise hazardous. Where contaminants are known to be present in quantities that are irritating or harmful to the occupants' health or are hazardous in a fire, both naturally and mechanically ventilated spaces must be equipped with mechanical exhaust systems capable of collecting and removing the contaminants.

This chapter contains requirements for the installation of exhaust systems, with an emphasis on the structural integrity of the systems and equipment involved and the overall impact of the systems on the fire safety performance of the building. It includes requirements for the exhaust of commercial kitchen grease- and smoke-laden air, hazardous fumes and toxic gases, clothes dryer moisture and heat and dust, stock and refuse materials.

Chapter 6 Duct Systems. Chapter 6 of the code regulates the materials and methods used for constructing and installing ducts, plenums, system controls, exhaust systems, fire protection systems and related components that affect the overall performance of a building's air distribution system and the reasonable protection of life and property from the hazards associated with air-moving equipment and systems. This chapter contains requirements for the installation of supply, return and exhaust air systems. Specific exhaust systems are also addressed in Chapter 5. Information on the design of duct systems is limited to that in Section 603.2. The code is very much concerned with the structural integrity of the systems and the overall impact of the systems on the fire safety and life safety performance of the building. Design considerations such as duct sizing, maximum efficiency, cost effectiveness, occupant comfort and convenience are the responsibility of the design professional. The provisions for the protection of duct penetrations of wall, floor, ceiling and roof assemblies are extracted from the *International Building Code.*

Chapter 7 Combustion Air. Complete combustion of solid and liquid fuel is essential for the proper operation of appliances, for control of harmful emissions and for achieving maximum fuel efficiency.

The specific combustion air requirements provided in previous editions of the code have been deleted in favor of a single section that directs the user to NFPA 31 for oil-fired appliance combustion air requirements and the manufacturer's installation instructions for solid-fuel burning appliances. For gas-fired appliances, the provisions of the *International Fuel Gas Code* are applicable.

Chapter 8 Chimneys and Vents. Chapter 8 is intended to regulate the design, construction, installation, maintenance, repair and approval of chimneys, vents and their connections to solid and liquid fuel-burning appliances. The requirements of this chapter are intended to achieve the complete removal of the products of combustion from fuel-burning appliances and equipment. This chapter includes regulations for the proper selection, design, construction and installation of a chimney or vent, along with appropriate measures to minimize the related potential fire hazards. A chimney or vent must be designed for the type of appliance or equipment it serves. Chimneys and vents are designed for specific applications depending on the flue gas temperatures and the type of fuel being burned in the appliance. Chimneys and vents for gas-fired appliances are covered in the *International Fuel Gas Code.*

Chapter 9 Specific Appliances, Fireplaces and Solid Fuel-burning Equipment. Chapter 9 sets minimum construction and performance criteria for fireplaces, appliances and equipment and provides for the safe installation of these items. It reflects the code's intent to specifically address all of the types of appliances that the code intends to regulate. Other regulations affecting the installation of solid fuel-burning fireplaces, appliances and accessory appliances are found in Chapters 3, 6, 7, 8, 10, 11, 12, 13 and 14.

Chapter 10 Boilers, Water Heaters and Pressure Vessels. Chapter 10 presents regulations for the proper installation of boilers, water heaters and pressure vessels to protect life and property from the hazards associated with those appliances and vessels. It applies to all types of boilers and pressure vessels, regardless of size, heat input, operating pressure or operating temperature.

Because pressure vessels are closed containers designed to contain liquids, gases or both under pressure, they must be designed and installed to prevent structural failures that can result in extremely hazardous situations. Certain safety features are therefore provided in Chapter 10 to reduce the potential for explosion hazards.

Chapter 11 Refrigeration. Chapter 11 contains regulations pertaining to the life safety of building occupants. These regulations establish minimum requirements to achieve the proper design, construction, installation and operation of refrigeration systems. Refrigeration systems are a combination of interconnected components and piping assembled to form a closed circuit in which a refrigerant is circulated. The system's function is to extract heat from a location or medium, and to reject that heat to a different location or medium. This chapter establishes reasonable safeguards for the occupants by defining and mandating practices that are consistent with the practices and experience of the industry.

Chapter 12 Hydronic Piping. Hydronic piping includes piping, fittings and valves used in building space conditioning systems. Applications include hot water, chilled water, steam, steam condensate, brines and water/antifreeze mixtures. Chapter 12 contains the provisions that govern the construction, installation, alteration and repair of all hydronic piping systems that affect reliability, serviceability, energy efficiency and safety.

Chapter 13 Fuel Oil Piping and Storage. Chapter 13 regulates the design and installation of fuel oil storage and piping systems. The regulations include reference to construction standards for above-ground and underground storage tanks, material standards for piping systems (both above-ground and underground) and extensive requirements for the proper assembly of system piping and components. The *International Fire Code* (IFC) covers subjects not addressed in detail here. The provisions in this chapter are intended to prevent fires, leaks and spills involving fuel oil storage and piping systems.

Chapter 14 Solar Systems. Chapter 14 establishes provisions for the safe installation, operation and repair of solar energy systems used for space heating or cooling, domestic hot water heating or processing. Although such systems use components similar to those of conventional mechanical equipment, many of these provisions are unique to solar energy systems.

Chapter 15 Referenced Standards. Chapter 15 lists all of the product and installation standards and codes that are referenced throughout Chapters 1 through 14. As stated in Section 102.8, these standards and codes become an enforceable part of the code (to the prescribed extent of the reference) as if printed in the body of the code. Chapter 15 provides the full title and edition year of the standards and codes in addition to the address of the promulgators and the section numbers in which the standards and codes are referenced.

Appendix A Chimney Connector Pass-throughs. Appendix A provides figures that illustrate various requirements in the body of the code. Figure A-1 illustrates the chimney connector clearance requirements of Table 803.10.4.

Appendix B Recommended Permit Fee Schedule. Appendix B provides a sample permit fee schedule for mechanical permits. The local jurisdiction can adopt this appendix and fill in the dollar amounts in the blank spaces to establish their official permit fee schedule. The ICC does not establish permit fees because the code is adopted throughout the country and there are vast differences in operating budgets between different parts of the country, as well as between large and small municipalities within the same region.

LEGISLATION

Jurisdictions wishing to adopt the 2015 *International Mechanical Code* as an enforceable regulation governing mechanical systems should ensure that certain factual information is included in the adopting legislation at the time adoption is being considered by the appropriate governmental body. The following sample adoption legislation addresses several key elements, including the information required for insertion into the code text.

SAMPLE LEGISLATION FOR ADOPTION OF THE *INTERNATIONAL MECHANICAL CODE* ORDINANCE NO._____

A**[N]** **[ORDINANCE/STATUTE/REGULATION]** of the **[JURISDICTION]** adopting the 2015 edition of the *International Mechanical Code*, regulating and governing the design, construction, quality of materials, erection, installation, alteration, repair, location, relocation, replacement, addition to, use or maintenance of mechanical systems in the **[JURISDICTION]**; providing for the issuance of permits and collection of fees therefor; repealing **[ORDINANCE/STATUTE/REGULATION]** No. _____ of the **[JURISDICTION]** and all other ordinances or parts of laws in conflict therewith.

The **[GOVERNING BODY]** of the **[JURISDICTION]** does ordain as follows:

Section 1. That a certain document, three (3) copies of which are on file in the office of the **[TITLE OF JURISDICTION'S KEEPER OF RECORDS]** of **[NAME OF JURISDICTION]**, being marked and designated as the *International Mechanical Code*, 2015 edition, including Appendix Chapters **[FILL IN THE APPENDIX CHAPTERS BEING ADOPTED]**, as published by the International Code Council, be and is hereby adopted as the Mechanical Code of the **[JURISDICTION]**, in the State of **[STATE NAME]** regulating and governing the design, construction, quality of materials, erection, installation, alteration, repair, location, relocation, replacement, addition to, use or maintenance of mechanical systems as herein provided; providing for the issuance of permits and collection of fees therefor; and each and all of the regulations, provisions, penalties, conditions and terms of said Mechanical Code on file in the office of the **[JURISDICTION]** are hereby referred to, adopted, and made a part hereof, as if fully set out in this legislation, with the additions, insertions, deletions and changes, if any, prescribed in Section 2 of this ordinance.

Section 2. The following sections are hereby revised:

Section 101.1. Insert: **[NAME OF JURISDICTION]**

Section 106.5.2. Insert: **[APPROPRIATE SCHEDULE]**

Section 106.5.3. Insert: **[PERCENTAGES IN TWO LOCATIONS]**

Section 108.4. Insert: **[OFFENSE, DOLLAR AMOUNT, NUMBER OF DAYS]**

Section 108.5. Insert: **[DOLLAR AMOUNT IN TWO LOCATIONS]**

Section 3. That **[ORDINANCE/STATUTE/REGULATION]** No. _____ of **[JURISDICTION]** entitled **[FILL IN HERE THE COMPLETE TITLE OF THE LEGISLATION OR LAWS IN EFFECT AT THE PRESENT TIME SO THAT THEY WILL BE REPEALED BY DEFINITE MENTION]** and all other ordinances or parts of laws in conflict herewith are hereby repealed.

Section 4. That if any section, subsection, sentence, clause or phrase of this legislation is, for any reason, held to be unconstitutional, such decision shall not affect the validity of the remaining portions of this ordinance. The **[GOVERNING BODY]** hereby declares that it would have passed this law, and each section, subsection, clause or phrase thereof, irrespective of the fact that any one or more sections, subsections, sentences, clauses and phrases be declared unconstitutional.

Section 5. That nothing in this legislation or in the Mechanical Code hereby adopted shall be construed to affect any suit or proceeding impending in any court, or any rights acquired, or liability incurred, or any cause or causes of action acquired or existing, under any act or ordinance hereby repealed as cited in Section 2 of this law; nor shall any just or legal right or remedy of any character be lost, impaired or affected by this legislation.

Section 6. That the **[JURISDICTION'S KEEPER OF RECORDS]** is hereby ordered and directed to cause this legislation to be published. (An additional provision may be required to direct the number of times the legislation is to be published and to specify that it is to be in a newspaper in general circulation. Posting may also be required.)

Section 7. That this law and the rules, regulations, provisions, requirements, orders and matters established and adopted hereby shall take effect and be in full force and effect **[TIME PERIOD]** from and after the date of its final passage and adoption.

TABLE OF CONTENTS

CHAPTER 1

SCOPE AND ADMINISTRATION

PART 1–SCOPE AND APPLICATION

SECTION 101
GENERAL

[A] 101.1 Title. These regulations shall be known as the *Mechanical Code* of **[NAME OF JURISDICTION]**, hereinafter referred to as "this code."

[A] 101.2 Scope. This code shall regulate the design, installation, maintenance, *alteration* and inspection of mechanical systems that are permanently installed and utilized to provide control of environmental conditions and related processes within buildings. This code shall also regulate those mechanical systems, system components, *equipment* and appliances specifically addressed herein. The installation of fuel gas distribution piping and *equipment*, fuel gas-fired appliances and fuel gas-fired *appliance* venting systems shall be regulated by the *International Fuel Gas Code*.

> **Exception:** Detached one- and two-family dwellings and multiple single-family dwellings (townhouses) not more than three stories high with separate means of egress and their accessory structures shall comply with the *International Residential Code*.

> **[A] 101.2.1 Appendices.** Provisions in the appendices shall not apply unless specifically adopted.

[A] 101.3 Intent. The purpose of this code is to establish minimum standards to provide a reasonable level of safety, health, property protection and public welfare by regulating and controlling the design, construction, installation, quality of materials, location, operation and maintenance or use of mechanical systems.

[A] 101.4 Severability. If a section, subsection, sentence, clause or phrase of this code is, for any reason, held to be unconstitutional, such decision shall not affect the validity of the remaining portions of this code.

SECTION 102
APPLICABILITY

[A] 102.1 General. Where there is a conflict between a general requirement and a specific requirement, the specific requirement shall govern. Where, in a specific case, different sections of this code specify different materials, methods of construction or other requirements, the most restrictive shall govern.

[A] 102.2 Existing installations. Except as otherwise provided for in this chapter, a provision in this code shall not require the removal, *alteration* or abandonment of, nor prevent the continued utilization and maintenance of, a mechanical system lawfully in existence at the time of the adoption of this code.

[A] 102.3 Maintenance. Mechanical systems, both existing and new, and parts thereof shall be maintained in proper operating condition in accordance with the original design and in a safe and sanitary condition. Devices or safeguards that are required by this code shall be maintained in compliance with the edition of the code under which they were installed. The owner or the owner's authorized agent shall be responsible for maintenance of mechanical systems. To determine compliance with this provision, the code official shall have the authority to require a mechanical system to be reinspected.

The inspection for maintenance of HVAC systems shall be performed in accordance with ASHRAE/ACCA/ANSI Standard 180.

[A] 102.4 Additions, alterations or repairs. Additions, alterations, renovations or repairs to a mechanical system shall conform to that required for a new mechanical system without requiring the existing mechanical system to comply with all of the requirements of this code. Additions, alterations or repairs shall not cause an existing mechanical system to become unsafe, hazardous or overloaded.

Minor additions, alterations, renovations and repairs to existing mechanical systems shall meet the provisions for new construction, unless such work is done in the same manner and arrangement as was in the existing system, is not hazardous and is *approved*.

[A] 102.5 Change in occupancy. It shall be unlawful to make a change in the *occupancy* of any structure which will subject the structure to any special provision of this code applicable to the new *occupancy* without approval. The code official shall certify that such structure meets the intent of the provisions of law governing building construction for the proposed new *occupancy* and that such change of *occupancy* does not result in any hazard to the public health, safety or welfare.

[A] 102.6 Historic buildings. The provisions of this code relating to the construction, *alteration*, repair, enlargement, restoration, relocation or moving of buildings or structures shall not be mandatory for existing buildings or structures identified and classified by the state or local jurisdiction as historic buildings where such buildings or structures are judged by the code official to be safe and in the public interest of health, safety and welfare regarding any proposed construction, *alteration*, repair, enlargement, restoration, relocation or moving of buildings.

[A] 102.7 Moved buildings. Except as determined by Section 102.2, mechanical systems that are a part of buildings or structures moved into or within the jurisdiction shall comply with the provisions of this code for new installations.

[A] 102.8 Referenced codes and standards. The codes and standards referenced herein shall be those that are listed in Chapter 15 and such codes and standards shall be considered as part of the requirements of this code to the prescribed extent of each such reference and as further regulated in Sections 102.8.1 and 102.8.2.

> **Exception:** Where enforcement of a code provision would violate the conditions of the listing of the *equipment* or *appliance*, the conditions of the listing and the manufacturer's installation instructions shall apply.

[A] 102.8.1 Conflicts. Where conflicts occur between provisions of this code and the referenced standards, the provisions of this code shall apply.

[A] 102.8.2 Provisions in referenced codes and standards. Where the extent of the reference to a referenced code or standard includes subject matter that is within the scope of this code, the provisions of this code, as applicable, shall take precedence over the provisions in the referenced code or standard.

[A] 102.9 Requirements not covered by this code. Requirements necessary for the strength, stability or proper operation of an existing or proposed mechanical system, or for the public safety, health and general welfare, not specifically covered by this code, shall be determined by the code official.

[A] 102.10 Other laws. The provisions of this code shall not be deemed to nullify any provisions of local, state or federal law.

[A] 102.11 Application of references. Reference to chapter section numbers, or to provisions not specifically identified by number, shall be construed to refer to such chapter, section or provision of this code.

PART 2–ADMINISTRATION AND ENFORCEMENT

SECTION 103
DEPARTMENT OF MECHANICAL INSPECTION

[A] 103.1 General. The department of mechanical inspection is hereby created and the executive official in charge thereof shall be known as the code official.

[A] 103.2 Appointment. The code official shall be appointed by the chief appointing authority of the jurisdiction.

[A] 103.3 Deputies. In accordance with the prescribed procedures of this jurisdiction and with the concurrence of the appointing authority, the code official shall have the authority to appoint a deputy code official, other related technical officers, inspectors and other employees. Such employees shall have powers as delegated by the code official.

[A] 103.4 Liability. The code official, member of the board of appeals or employee charged with the enforcement of this code, while acting for the jurisdiction in good faith and without malice in the discharge of the duties required by this code or other pertinent law or ordinance, shall not thereby be rendered civilly or criminally liable personally, and is hereby relieved from personal liability for any damage accruing to persons or property as a result of an act or by reason of an act or omission in the discharge of official duties.

[A] 103.4.1 Legal defense. Any suit or criminal complaint instituted against any officer or employee because of an act performed by that officer or employee in the lawful discharge of duties and under the provisions of this code shall be defended by the legal representatives of the jurisdiction until the final termination of the proceedings. The code official or any subordinate shall not be liable for costs in an action, suit or proceeding that is instituted in pursuance of the provisions of this code.

SECTION 104
DUTIES AND POWERS OF THE CODE OFFICIAL

[A] 104.1 General. The code official is hereby authorized and directed to enforce the provisions of this code. The code official shall have the authority to render interpretations of this code and to adopt policies and procedures in order to clarify the application of its provisions. Such interpretations, policies and procedures shall be in compliance with the intent and purpose of this code. Such policies and procedures shall not have the effect of waiving requirements specifically provided for in this code.

[A] 104.2 Applications and permits. The code official shall receive applications, review *construction documents* and issue permits for the installation and *alteration* of mechanical systems, inspect the premises for which such permits have been issued and enforce compliance with the provisions of this code.

[A] 104.3 Inspections. The code official shall make all of the required inspections, or shall accept reports of inspection by *approved* agencies or individuals. Reports of such inspections shall be in writing and be certified by a responsible officer of such *approved* agency or by the responsible individual. The code official is authorized to engage such expert opinion as deemed necessary to report upon unusual technical issues that arise, subject to the approval of the appointing authority.

[A] 104.4 Right of entry. Where it is necessary to make an inspection to enforce the provisions of this code, or where the code official has reasonable cause to believe that there exists in a building or upon any premises any conditions or violations of this code that make the building or premises unsafe, insanitary, dangerous or hazardous, the code official shall have the authority to enter the building or premises at all reasonable times to inspect or to perform the duties imposed upon the code official by this code. If such building or premises is occupied, the code official shall present credentials to the occupant and request entry. If such building or premises is unoccupied, the code official shall first make a reasonable effort to locate the owner, the owner's authorized agent or other person having charge or control of the building or premises and request entry. If entry is refused, the code official has recourse to every remedy provided by law to secure entry.

Where the code official has first obtained a proper inspection warrant or other remedy provided by law to secure entry, an owner, the owner's authorized agent or occupant or person having charge, care or control of the building or premises shall not fail or neglect, after proper request is made as herein provided, to promptly permit entry therein by the code official for the purpose of inspection and examination pursuant to this code.

[A] 104.5 Identification. The code official shall carry proper identification when inspecting structures or premises in the performance of duties under this code.

[A] 104.6 Notices and orders. The code official shall issue all necessary notices or orders to ensure compliance with this code.

[A] 104.7 Department records. The code official shall keep official records of applications received, permits and certificates issued, fees collected, reports of inspections, and notices and orders issued. Such records shall be retained in the official records for the period required for retention of public records.

SECTION 105
APPROVAL

[A] 105.1 Modifications. Where there are practical difficulties involved in carrying out the provisions of this code, the code official shall have the authority to grant modifications for individual cases upon application of the owner or owner's authorized agent, provided that the code official shall first find that special individual reason makes the strict letter of this code impractical and the modification is in compliance with the intent and purpose of this code and does not lessen health, life and fire safety requirements. The details of action granting modifications shall be recorded and entered in the files of the mechanical inspection department.

[A] 105.2 Alternative materials, methods, equipment and appliances. The provisions of this code are not intended to prevent the installation of any material or to prohibit any method of construction not specifically prescribed by this code, provided that any such alternative has been *approved*. An alternative material or method of construction shall be *approved* where the code official finds that the proposed design is satisfactory and complies with the intent of the provisions of this code, and that the material, method or work offered is, for the purpose intended, not less than the equivalent of that prescribed in this code in quality, strength, effectiveness, fire resistance, durability and safety. Where the alternative material, design or method of construction is not approved, the *code official* shall respond in writing, stating the reasons the alternative was not approved.

[A] 105.2.1 Research reports. Supporting data, where necessary to assist in the approval of materials or assemblies not specifically provided for in this code, shall consist of valid research reports from *approved* sources.

[A] 105.3 Required testing. Where there is insufficient evidence of compliance with the provisions of this code, or evidence that a material or method does not conform to the requirements of this code, or in order to substantiate claims for alternative materials or methods, the code official shall have the authority to require tests as evidence of compliance to be made at no expense to the jurisdiction.

[A] 105.3.1 Test methods. Test methods shall be as specified in this code or by other recognized test standards. In the absence of recognized and accepted test methods, the code official shall approve the testing procedures.

[A] 105.3.2 Testing agency. Tests shall be performed by an *approved* agency.

[A] 105.3.3 Test reports. Reports of tests shall be retained by the code official for the period required for retention of public records.

[A] 105.4 Approved materials and equipment. Materials, *equipment* and devices *approved* by the code official shall be constructed and installed in accordance with such approval.

[A] 105.5 Material, equipment and appliance reuse. Materials, *equipment*, appliances and devices shall not be reused unless such elements have been reconditioned, tested and placed in good and proper working condition and *approved*.

SECTION 106
PERMITS

[A] 106.1 Where required. An owner, owner's authorized agent or contractor who desires to erect, install, enlarge, alter, repair, remove, convert or replace a mechanical system, the installation of which is regulated by this code, or to cause such work to be performed, shall first make application to the code official and obtain the required permit for the work.

Exception: Where *equipment* and *appliance* replacements or repairs must be performed in an emergency situation, the permit application shall be submitted within the next working business day of the department of mechanical inspection.

[A] 106.1.1 Annual permit. Instead of an individual construction permit for each alteration to an already *approved* system or equipment or application installation, the code official is authorized to issue an annual permit upon application therefor to any person, firm or corporation regularly employing one or more qualified tradespersons in the building, structure or on the premises owned or operated by the applicant for the permit.

[A] 106.1.2 Annual permit records. The person to whom an annual permit is issued shall keep a detailed record of alterations made under such annual permit. The code official shall have access to such records at all times or such records shall be filed with the code official as designated.

[A] 106.2 Permits not required. Permits shall not be required for the following:

1. Portable heating appliances.

2. Portable ventilation appliances and *equipment*.

3. Portable cooling units.

4. Steam, hot water or chilled water piping within any heating or cooling *equipment* or appliances regulated by this code.

5. The replacement of any minor part that does not alter the approval of *equipment* or an *appliance* or make such *equipment* or *appliance* unsafe.

6. Portable evaporative coolers.

7. Self-contained refrigeration systems that contain 10 pounds (4.5 kg) or less of refrigerant, or that are actuated by motors of 1 horsepower (0.75 kW) or less.

8. Portable fuel cell appliances that are not connected to a fixed piping system and are not interconnected to a power grid.

Exemption from the permit requirements of this code shall not be deemed to grant authorization for work to be done in violation of the provisions of this code or other laws or ordinances of this jurisdiction.

[A] 106.3 Application for permit. Each application for a permit, with the required fee, shall be filed with the code official on a form furnished for that purpose and shall contain a general description of the proposed work and its location. The application shall be signed by the owner or the owner's authorized agent. The permit application shall indicate the proposed *occupancy* of all parts of the building and of that portion of the site or lot, if any, not covered by the building or structure and shall contain such other information required by the code official.

[A] 106.3.1 Construction documents. *Construction documents*, engineering calculations, diagrams and other data shall be submitted in two or more sets with each application for a permit. The code official shall require *construction documents*, computations and specifications to be prepared and designed by a *registered design professional* where required by state law. Where special conditions exist, the code official is authorized to require additional *construction documents* to be prepared by a *registered design professional*. *Construction documents* shall be drawn to scale and shall be of sufficient clarity to indicate the location, nature and extent of the work proposed and show in detail that the work conforms to the provisions of this code. *Construction documents* for buildings more than two stories in height shall indicate where penetrations will be made for mechanical systems, and the materials and methods for maintaining required structural safety, fire-resistance rating and fireblocking.

Exception: The code official shall have the authority to waive the submission of *construction documents*, calculations or other data if the nature of the work applied for is such that reviewing of *construction documents* is not necessary to determine compliance with this code.

[A] 106.3.2 Preliminary inspection. Before a permit is issued, the code official is authorized to inspect and evaluate the systems, *equipment*, buildings, devices, premises and spaces or areas to be used.

[A] 106.3.3 Time limitation of application. An application for a permit for any proposed work shall be deemed to have been abandoned 180 days after the date of filing, unless such application has been pursued in good faith or a permit has been issued; except that the code official shall have the authority to grant one or more extensions of time for additional periods not exceeding 180 days each. The extension shall be requested in writing and justifiable cause demonstrated.

[A] 106.4 Permit issuance. The application, *construction documents* and other data filed by an applicant for a permit shall be reviewed by the code official. If the code official finds that the proposed work conforms to the requirements of this code and all laws and ordinances applicable thereto, and that the fees specified in Section 106.5 have been paid, a permit shall be issued to the applicant.

[A] 106.4.1 Approved construction documents. When the code official issues the permit where *construction documents* are required, the *construction documents* shall be endorsed in writing and stamped "*APPROVED.*" Such *approved construction documents* shall not be changed, modified or altered without authorization from the code official. Work shall be done in accordance with the *approved construction documents*.

The code official shall have the authority to issue a permit for the construction of part of a mechanical system before the *construction documents* for the entire system have been submitted or *approved*, provided adequate information and detailed statements have been filed complying with all pertinent requirements of this code. The holder of such permit shall proceed at his or her own risk without assurance that the permit for the entire mechanical system will be granted.

[A] 106.4.2 Validity. The issuance of a permit or approval of *construction documents* shall not be construed to be a permit for, or an approval of, any violation of any of the provisions of this code or of other ordinances of the jurisdiction. A permit presuming to give authority to violate or cancel the provisions of this code shall be invalid.

The issuance of a permit based upon *construction documents* and other data shall not prevent the code official from thereafter requiring the correction of errors in said *construction documents* and other data or from preventing building operations from being carried on thereunder where in violation of this code or of other ordinances of this jurisdiction.

[A] 106.4.3 Expiration. Every permit issued by the code official under the provisions of this code shall expire by limitation and become null and void if the work authorized by such permit is not commenced within 180 days from the date of such permit, or if the work authorized by such permit is suspended or abandoned at any time after the work is commenced for a period of 180 days. Before such work recommences, a new permit shall be first obtained and the fee therefor shall be one-half the amount required for a new permit for such work, provided that changes have not been made and will not be made in the original *construction documents* for such work, and provided further that such suspension or abandonment has not exceeded one year.

[A] 106.4.4 Extensions. A permittee holding an unexpired permit shall have the right to apply for an extension of the time within which the permittee will commence work under that permit when work is unable to be commenced within the time required by this section for good and satisfactory reasons. The code official shall extend the time for action by the permittee for a period not exceeding 180 days if there is reasonable cause. A permit shall not be extended more than once. The fee for an extension shall be one-half the amount required for a new permit for such work.

[A] 106.4.5 Suspension or revocation of permit. The code official shall have the authority to suspend or revoke a permit issued under the provisions of this code wherever the permit is issued in error or on the basis of incorrect, inaccurate or incomplete information, or in violation of any ordinance or regulation or any of the provisions of this code.

[A] 106.4.6 Retention of construction documents. One set of *approved construction documents* shall be retained by the code official for a period of not less than 180 days from date of completion of the permitted work, or as required by state or local laws. One set of *approved construction documents* shall be returned to the applicant, and said set shall be kept on the site of the building or job at all times during which the work authorized thereby is in progress.

[A] 106.4.7 Previous approvals. This code shall not require changes in the *construction documents*, construction or designated *occupancy* of a structure for which a lawful permit has been heretofore issued or otherwise lawfully authorized, and the construction of which has been pursued in good faith within 180 days after the effective date of this code and has not been abandoned.

[A] 106.4.8 Posting of permit. The permit or a copy shall be kept on the site of the work until the completion of the project.

[A] 106.5 Fees. A permit shall not be issued until the fees prescribed in Section 106.5.2 have been paid, nor shall an amendment to a permit be released until the additional fee, if any, due to an increase of the mechanical system, has been paid.

[A] 106.5.1 Work commencing before permit issuance. Any person who commences work on a mechanical system before obtaining the necessary permits shall be subject to 100 percent of the usual permit fee in addition to the required permit fees.

[A] 106.5.2 Fee schedule. The fees for mechanical work shall be as indicated in the following schedule.

**[JURISDICTION TO INSERT
APPROPRIATE SCHEDULE]**

[A] 106.5.3 Fee refunds. The code official shall authorize the refunding of fees as follows.

1. The full amount of any fee paid hereunder which was erroneously paid or collected.

2. Not more than **[SPECIFY PERCENTAGE]** percent of the permit fee paid where work has not been done under a permit issued in accordance with this code.

3. Not more than **[SPECIFY PERCENTAGE]** percent of the plan review fee paid where an application for a permit for which a plan review fee has been paid is withdrawn or canceled before any plan review effort has been expended.

The code official shall not authorize the refunding of any fee paid, except upon written application filed by the original permittee not later than 180 days after the date of fee payment.

SECTION 107
INSPECTIONS AND TESTING

[A] 107.1 General. The code official is authorized to conduct such inspections as are deemed necessary to determine compliance with the provisions of this code. Construction or work for which a permit is required shall be subject to inspection by the code official, and such construction or work shall remain accessible and exposed for inspection purposes until *approved*. Approval as a result of an inspection shall not be construed to be an approval of a violation of the provisions of this code or of other ordinances of the jurisdiction. Inspections presuming to give authority to violate or cancel the provisions of this code or of other ordinances of the jurisdiction shall not be valid.

[A] 107.2 Required inspections and testing. The code official, upon notification from the permit holder or the permit holder's agent, shall make the following inspections and other such inspections as necessary, and shall either release that portion of the construction or shall notify the permit holder or the permit holder's agent of violations that must be corrected. The holder of the permit shall be responsible for the scheduling of such inspections.

1. Underground inspection shall be made after trenches or ditches are excavated and bedded, piping installed, and before backfill is put in place. Where excavated soil contains rocks, broken concrete, frozen chunks and other rubble that would damage or break the piping or cause corrosive action, clean backfill shall be on the job site.

2. Rough-in inspection shall be made after the roof, framing, fireblocking and bracing are in place and all ducting and other components to be concealed are complete, and prior to the installation of wall or ceiling membranes.

3. Final inspection shall be made upon completion of the mechanical system.

Exception: Ground-source heat pump loop systems tested in accordance with Section 1210.10 shall be permitted to be backfilled prior to inspection.

The requirements of this section shall not be considered to prohibit the operation of any heating *equipment* or appliances installed to replace existing heating *equipment* or appliances serving an occupied portion of a structure provided that a request for inspection of such heating *equipment* or appliances has been filed with the department not more than 48 hours after such replacement work is completed, and before any portion of such *equipment* or appliances is concealed by any permanent portion of the structure.

[A] 107.2.1 Other inspections. In addition to the inspections specified in Section 107.2, the code official is authorized to make or require other inspections of any construction work to ascertain compliance with the provisions of this code and other laws that are enforced.

[A] 107.2.2 Inspection requests. It shall be the duty of the holder of the permit or their duly authorized agent to notify the code official when work is ready for inspection. It shall be the duty of the permit holder to provide *access* to and means for inspections of such work that are required by this code.

[A] 107.2.3 Approval required. Work shall not be done beyond the point indicated in each successive inspection without first obtaining the approval of the code official. The code official, upon notification, shall make the requested inspections and shall either indicate the portion of the construction that is satisfactory as completed, or notify the permit holder or his or her agent wherein the same fails to comply with this code. Any portions that do not comply shall be corrected and such portion shall not be covered or concealed until authorized by the code official.

[A] 107.2.4 Approved inspection agencies. The code official is authorized to accept reports of *approved* agencies, provided that such agencies satisfy the requirements as to qualifications and reliability.

[A] 107.2.5 Evaluation and follow-up inspection services. Prior to the approval of a prefabricated construction assembly having concealed mechanical work and the issuance of a mechanical permit, the code official shall require the submittal of an evaluation report on each prefabricated construction assembly, indicating the complete details of the mechanical system, including a description of the system and its components, the basis upon which the system is being evaluated, test results and similar information, and other data as necessary for the code official to determine conformance to this code.

[A] 107.2.5.1 Evaluation service. The code official shall designate the evaluation service of an *approved* agency as the evaluation agency, and review such agency's evaluation report for adequacy and conformance to this code.

[A] 107.2.5.2 Follow-up inspection. Except where ready access is provided to mechanical systems, service *equipment* and accessories for complete inspection at the site without disassembly or dismantling, the code official shall conduct the in-plant inspections as frequently as necessary to ensure conformance to the *approved* evaluation report or shall designate an independent, *approved* inspection agency to conduct such inspections. The inspection agency shall furnish the code official with the follow-up inspection manual and a report of inspections upon request, and the mechanical system shall have an identifying label permanently affixed to the system indicating that factory inspections have been performed.

[A] 107.2.5.3 Test and inspection records. Required test and inspection records shall be available to the code official at all times during the fabrication of the mechanical system and the erection of the building; or such records as the code official designates shall be filed.

[A] 107.3 Testing. Mechanical systems shall be tested as required in this code and in accordance with Sections 107.3.1 through 107.3.3. Tests shall be made by the permit holder and observed by the code official.

[A] 107.3.1 New, altered, extended or repaired systems. New mechanical systems and parts of existing systems, which have been altered, extended, renovated or repaired, shall be tested as prescribed herein to disclose leaks and defects.

[A] 107.3.2 Apparatus, material and labor for tests. Apparatus, material and labor required for testing a mechanical system or part thereof shall be furnished by the permit holder.

[A] 107.3.3 Reinspection and testing. Where any work or installation does not pass an initial test or inspection, the necessary corrections shall be made so as to achieve compliance with this code. The work or installation shall then be resubmitted to the code official for inspection and testing.

[A] 107.4 Approval. After the prescribed tests and inspections indicate that the work complies in all respects with this code, a notice of approval shall be issued by the code official.

[A] 107.4.1 Revocation. The code official is authorized to, in writing, suspend or revoke a notice of approval issued under the provisions of this code wherever the notice is issued in error, on the basis of incorrect information supplied, or where it is determined that the building or structure, premise or portion thereof is in violation of any ordinance or regulation or any of the provisions of this code.

[A] 107.5 Temporary connection. The code official shall have the authority to authorize the temporary connection of a mechanical system to the sources of energy for the purpose of testing mechanical systems or for use under a temporary certificate of *occupancy*.

[A] 107.6 Connection of service utilities. A person shall not make connections from a utility, source of energy, fuel or power to any building or system that is regulated by this code for which a permit is required, until authorized by the code official.

SECTION 108
VIOLATIONS

[A] 108.1 Unlawful acts. It shall be unlawful for a person, firm or corporation to erect, construct, alter, repair, remove, demolish or utilize a mechanical system, or cause same to be done, in conflict with or in violation of any of the provisions of this code.

[A] 108.2 Notice of violation. The code official shall serve a notice of violation or order to the person responsible for the erection, installation, *alteration*, extension, repair, removal or demolition of mechanical work in violation of the provisions of this code, or in violation of a detail statement or the *approved construction documents* thereunder, or in violation of a permit or certificate issued under the provisions of this code. Such order shall direct the discontinuance of the illegal action or condition and the abatement of the violation.

[A] 108.3 Prosecution of violation. If the notice of violation is not complied with promptly, the code official shall request the legal counsel of the jurisdiction to institute the appropriate proceeding at law or in equity to restrain, correct or abate such violation, or to require the removal or termination of the unlawful *occupancy* of the structure in violation of the provisions of this code or of the order or direction made pursuant thereto.

[A] 108.4 Violation penalties. Persons who shall violate a provision of this code or shall fail to comply with any of the requirements thereof or who shall erect, install, alter or repair mechanical work in violation of the *approved construction documents* or directive of the code official, or of a permit or certificate issued under the provisions of this code, shall be guilty of a **[SPECIFY OFFENSE]**, punishable by a fine of not more than **[AMOUNT]** dollars or by imprisonment not exceeding **[NUMBER OF DAYS]**, or both such fine and imprisonment. Each day that a violation continues after due notice has been served shall be deemed a separate offense.

[A] 108.5 Stop work orders. Upon notice from the code official that mechanical work is being performed contrary to the provisions of this code or in a dangerous or unsafe manner, such work shall immediately cease. Such notice shall be in writing and shall be given to the owner of the property, or to the owner's authorized agent, or to the person doing the work. The notice shall state the conditions under which work is authorized to resume. Where an emergency exists, the code official shall not be required to give a written notice prior to stopping the work. Any person who shall continue any work on the system after having been served with a stop work order, except such work as that person is directed to perform to remove a violation or unsafe condition, shall be liable for a fine of not less than **[AMOUNT]** dollars or more than **[AMOUNT]** dollars.

[A] 108.6 Abatement of violation. The imposition of the penalties herein prescribed shall not preclude the legal officer of the jurisdiction from instituting appropriate action to prevent unlawful construction or to restrain, correct or abate a violation, or to prevent illegal *occupancy* of a building, structure or premises, or to stop an illegal act, conduct, business or utilization of the mechanical system on or about any premises.

[A] 108.7 Unsafe mechanical systems. A mechanical system that is unsafe, constitutes a fire or health hazard, or is otherwise dangerous to human life, as regulated by this code, is hereby declared as an unsafe mechanical system. Use of a mechanical system regulated by this code constituting a hazard to health, safety or welfare by reason of inadequate maintenance, dilapidation, fire hazard, disaster, damage or abandonment is hereby declared an unsafe use. Such unsafe *equipment* and appliances are hereby declared to be a public nuisance and shall be abated by repair, rehabilitation, demolition or removal.

[A] 108.7.1 Authority to condemn mechanical systems. Whenever the code official determines that any mechanical system, or portion thereof, regulated by this code has become hazardous to life, health, property, or has become insanitary, the code official shall order in writing that such system either be removed or restored to a safe condition. A time limit for compliance with such order shall be specified in the written notice. A person shall not use or maintain a defective mechanical system after receiving such notice.

Where such mechanical system is to be disconnected, written notice as prescribed in Section 108.2 shall be given. In cases of immediate danger to life or property, such disconnection shall be made immediately without such notice.

[A] 108.7.2 Authority to order disconnection of energy sources. The code official shall have the authority to order disconnection of energy sources supplied to a building, structure or mechanical system regulated by this code, where it is determined that the mechanical system or any portion thereof has become hazardous or unsafe. Written notice of such order to disconnect service and the causes therefor shall be given within 24 hours to the owner, the owner's authorized agent and occupant of such building, structure or premises, provided, however, that in cases of immediate danger to life or property, such disconnection shall be made immediately without such notice. Where energy sources are provided by a public utility, the code official shall immediately notify the serving utility in writing of the issuance of such order to disconnect.

[A] 108.7.3 Connection after order to disconnect. A person shall not make energy source connections to mechanical systems regulated by this code which have been disconnected or ordered to be disconnected by the code official, or the use of which has been ordered to be discontinued by the code official until the code official authorizes the reconnection and use of such mechanical systems.

Where a mechanical system is maintained in violation of this code, and in violation of a notice issued pursuant to the provisions of this section, the code official shall institute appropriate action to prevent, restrain, correct or abate the violation.

SECTION 109
MEANS OF APPEAL

[A] 109.1 Application for appeal. A person shall have the right to appeal a decision of the code official to the board of appeals. An application for appeal shall be based on a claim that the true intent of this code or the rules legally adopted thereunder have been incorrectly interpreted, the provisions of this code do not fully apply, or an equally good or better form of construction is proposed. The application shall be filed on a form obtained from the code official within 20 days after the notice was served.

[A] 109.1.1 Limitation of authority. The board of appeals shall not have authority relative to interpretation of the administration of this code nor shall such board be empowered to waive requirements of this code.

[A] 109.2 Membership of board. The board of appeals shall consist of five members appointed by the chief appointing authority as follows: one for 5 years; one for 4 years; one for 3 years; one for 2 years; and one for 1 year. Thereafter, each new member shall serve for 5 years or until a successor has been appointed.

[A] 109.2.1 Qualifications. The board of appeals shall consist of five individuals, one from each of the following professions or disciplines.

1. *Registered design professional* who is a registered architect; or a builder or superintendent of building construction with not less than 10 years' experience, 5 of which shall have been in responsible charge of work.

2. *Registered design professional* with structural engineering or architectural experience.

3. *Registered design professional* with mechanical and plumbing engineering experience; or a mechanical contractor with not less than 10 years' experience, 5 of which shall have been in responsible charge of work.

4. *Registered design professional* with electrical engineering experience; or an electrical contractor with not less than 10 years' experience, 5 of which shall have been in responsible charge of work.

5. *Registered design professional* with fire protection engineering experience; or a fire protection contractor with not less than 10 years' experience, 5 of which shall have been in responsible charge of work.

[A] 109.2.2 Alternate members. The chief appointing authority shall appoint two alternate members who shall be called by the board chairman to hear appeals during the absence or disqualification of a member. Alternate members shall possess the qualifications required for board membership and shall be appointed for 5 years, or until a successor has been appointed.

[A] 109.2.3 Chairman. The board shall annually select one of its members to serve as chairman.

[A] 109.2.4 Disqualification of member. A member shall not hear an appeal in which that member has a personal, professional or financial interest.

[A] 109.2.5 Secretary. The chief administrative officer shall designate a qualified clerk to serve as secretary to the board. The secretary shall file a detailed record of all proceedings in the office of the chief administrative officer.

[A] 109.2.6 Compensation of members. Compensation of members shall be determined by law.

[A] 109.3 Notice of meeting. The board shall meet upon notice from the chairman, within 10 days of the filing of an appeal, or at stated periodic meetings.

[A] 109.4 Open hearing. Hearings before the board shall be open to the public. The appellant, the appellant's representative, the code official and any person whose interests are affected shall be given an opportunity to be heard.

[A] 109.4.1 Procedure. The board shall adopt and make available to the public through the secretary procedures under which a hearing will be conducted. The procedures shall not require compliance with strict rules of evidence, but shall mandate that only relevant information be received.

[A] 109.5 Postponed hearing. When five members are not present to hear an appeal, either the appellant or the appellant's representative shall have the right to request a postponement of the hearing.

[A] 109.6 Board decision. The board shall modify or reverse the decision of the code official by a concurring vote of three members.

[A] 109.6.1 Resolution. The decision of the board shall be by resolution. Certified copies shall be furnished to the appellant and to the code official.

[A] 109.6.2 Administration. The code official shall take immediate action in accordance with the decision of the board.

[A] 109.7 Court review. Any person, whether or not a previous party of the appeal, shall have the right to apply to the appropriate court for a writ of certiorari to correct errors of law. Application for review shall be made in the manner and time required by law following the filing of the decision in the office of the chief administrative officer.

SECTION 110
TEMPORARY EQUIPMENT, SYSTEMS AND USES

[A] 110.1 General. The code official is authorized to issue a permit for temporary *equipment*, systems and uses. Such permits shall be limited as to time of service, but shall not be permitted for more than 180 days. The code official is authorized to grant extensions for demonstrated cause.

[A] 110.2 Conformance. Temporary *equipment*, systems and uses shall conform to the structural strength, fire safety, means of egress, accessibility, light, ventilation and sanitary requirements of this code as necessary to ensure the public health, safety and general welfare.

[A] 110.3 Temporary utilities. The code official is authorized to give permission to temporarily supply utilities before an installation has been fully completed and the final certificate of completion has been issued. The part covered by the temporary certificate shall comply with the requirements specified for temporary lighting, heat or power in the code.

[A] 110.4 Termination of approval. The code official is authorized to terminate such permit for temporary *equipment*, systems or uses and to order the temporary *equipment*, systems or uses to be discontinued.

CHAPTER 2

DEFINITIONS

SECTION 201
GENERAL

201.1 Scope. Unless otherwise expressly stated, the following words and terms shall, for the purposes of this code, have the meanings indicated in this chapter.

201.2 Interchangeability. Words used in the present tense include the future; words in the masculine gender include the feminine and neuter; the singular number includes the plural and the plural, the singular.

201.3 Terms defined in other codes. Where terms are not defined in this code and are defined in the *International Building Code, International Fire Code, International Fuel Gas Code* or the *International Plumbing Code,* such terms shall have meanings ascribed to them as in those codes.

201.4 Terms not defined. Where terms are not defined through the methods authorized by this section, such terms shall have ordinarily accepted meanings such as the context implies.

SECTION 202
GENERAL DEFINITIONS

ABRASIVE MATERIALS. Moderately abrasive particulate in high concentrations, and highly abrasive particulate in moderate and high concentrations, such as alumina, bauxite, iron silicate, sand and slag.

ABSORPTION SYSTEM. A refrigerating system in which refrigerant is pressurized by pumping a chemical solution of refrigerant in absorbent, and then separated by the addition of heat in a generator, condensed (to reject heat), expanded, evaporated (to provide refrigeration), and reabsorbed in an absorber to repeat the cycle; the system may be single or multiple effect, the latter using multiple stages or internally cascaded use of heat to improve efficiency.

ACCESS (TO). That which enables a device, *appliance* or *equipment* to be reached by ready access or by a means that first requires the removal or movement of a panel, door or similar obstruction [see also "Ready access (to)"].

AIR. All air supplied to mechanical *equipment* and appliances for *combustion,* ventilation, cooling and similar purposes. Standard air is air at standard temperature and pressure, namely, 70°F (21°C) and 29.92 inches of mercury (101.3 kPa).

AIR CONDITIONING. The treatment of air so as to control simultaneously the temperature, humidity, cleanness and distribution of the air to meet the requirements of a conditioned space.

AIR-CONDITIONING SYSTEM. A system that consists of heat exchangers, blowers, filters, supply, exhaust and return ducts, and shall include any apparatus installed in connection therewith.

AIR DISPERSION SYSTEM. Any diffuser system designed to both convey air within a room, space or area and diffuse air into that space while operating under positive pressure. Systems are commonly constructed of, but not limited to, fabric or plastic film.

AIR DISTRIBUTION SYSTEM. Any system of ducts, plenums and air-handling *equipment* that circulates air within a space or spaces and includes systems made up of one or more air-handling units.

AIR, EXHAUST. Air being removed from any space, *appliance* or piece of *equipment* and conveyed directly to the atmosphere by means of openings or ducts.

AIR-HANDLING UNIT. A blower or fan used for the purpose of distributing supply air to a room, space or area.

AIR, MAKEUP. Any combination of outdoor and transfer air intended to replace exhaust air and exfiltration.

AIR, OUTDOOR. Ambient air that enters a building through a ventilation system, through intentional openings for natural ventilation, or by infiltration.

AIR, TRANSFER. Air moved from one indoor space to another.

[A] ALTERATION. A change in a mechanical system that involves an extension, addition or change to the arrangement, type or purpose of the original installation.

APPLIANCE. A device or apparatus that is manufactured and designed to utilize energy and for which this code provides specific requirements.

APPLIANCE, EXISTING. Any *appliance* regulated by this code which was legally installed prior to the effective date of this code, or for which a permit to install has been issued.

APPLIANCE TYPE.

> **High-heat appliance.** Any *appliance* in which the products of *combustion* at the point of entrance to the flue under normal operating conditions have a temperature greater than 2,000°F (1093°C).

> **Low-heat appliance (residential appliance).** Any *appliance* in which the products of *combustion* at the point of entrance to the flue under normal operating conditions have a temperature of 1,000°F (538°C) or less.

> **Medium-heat appliance.** Any *appliance* in which the products of *combustion* at the point of entrance to the flue under normal operating conditions have a temperature of more than 1,000°F (538°C), but not greater than 2,000°F (1093°C).

APPLIANCE, VENTED. An *appliance* designed and installed in such a manner that all of the products of *combustion* are conveyed directly from the *appliance* to the outdoor atmosphere through an *approved chimney* or vent system.

[A] APPROVED. Acceptable to the code official.

[A] APPROVED AGENCY. An established and recognized agency that is regularly engaged in conducting tests or furnishing inspection services where such agency has been approved by the code official.

AUTOMATIC BOILER. Any class of boiler that is equipped with the controls and limit devices specified in Chapter 10.

BATHROOM. A room containing a bathtub, shower, spa or similar bathing fixture.

BOILER. A closed heating *appliance* intended to supply hot water or steam for space heating, processing or power purposes. Low-pressure boilers operate at pressures less than or equal to 15 pounds per square inch (psi) (103 kPa) for steam and 160 psi (1103 kPa) for water. High-pressure boilers operate at pressures exceeding those pressures.

BOILER ROOM. A room primarily utilized for the installation of a boiler.

BRAZED JOINT. A gas-tight joint obtained by the joining of metal parts with metallic mixtures or alloys which melt at a temperature above 1,000°F (538°C), but lower than the melting temperature of the parts to be joined.

BRAZING. A metal joining process wherein coalescence is produced by the use of a nonferrous filler metal having a melting point above 1,000°F (538°C), but lower than that of the base metal being joined. The filler material is distributed between the closely fitted surfaces of the joint by capillary attraction.

BREATHING ZONE. The region within an occupied space between planes 3 and 72 inches (76 and 1829 mm) above the floor and more than 2 feet (610 mm) from the walls of the space or from fixed air-conditioning *equipment*.

BTU. Abbreviation for British thermal unit, which is the quantity of heat required to raise the temperature of 1 pound (454 g) of water 1°F (0.56°C) (1 Btu = 1055 J).

[A] BUILDING. Any structure occupied or intended for supporting or sheltering any *occupancy*.

[BF] CEILING RADIATION DAMPER. A *listed* device installed in a ceiling membrane of a fire-resistance-rated floor/ceiling or roof/ceiling assembly to limit automatically the radiative heat transfer through an air inlet/outlet opening.

CHIMNEY. A primarily vertical structure containing one or more flues, for the purpose of carrying gaseous products of *combustion* and air from a fuel-burning *appliance* to the outdoor atmosphere.

> **Factory-built chimney.** A *listed* and *labeled chimney* composed of factory-made components, assembled in the field in accordance with manufacturer's instructions and the conditions of the listing.

> **Masonry chimney.** A field-constructed *chimney* composed of solid masonry units, bricks, stones or concrete.

> **Metal chimney.** A field-constructed *chimney* of metal.

CHIMNEY CONNECTOR. A pipe that connects a fuel-burning *appliance* to a *chimney*.

CLEARANCE. The minimum distance through air measured between the heat-producing surface of the mechanical *appliance*, device or *equipment* and the surface of the combustible material or assembly.

CLOSED COMBUSTION SOLID-FUEL-BURNING APPLIANCE. A heat-producing *appliance* that employs a *combustion* chamber that does not have openings other than the flue collar, fuel charging door and adjustable openings provided to control the amount of *combustion air* that enters the *combustion* chamber.

CLOTHES DRYER. An *appliance* used to dry wet laundry by means of heat.

[A] CODE. These regulations, subsequent amendments thereto, or any emergency rule or regulation that the administrative authority having jurisdiction has lawfully adopted.

[A] CODE OFFICIAL. The officer or other designated authority charged with the administration and enforcement of this code, or a duly authorized representative.

[BF] COMBINATION FIRE/SMOKE DAMPER. A *listed* device installed in ducts and air transfer openings designed to close automatically upon the detection of heat and resist the passage of flame and smoke. The device is installed to operate automatically, be controlled by a smoke detection system, and where required, is capable of being positioned from a fire command center.

COMBUSTIBLE ASSEMBLY. Wall, floor, ceiling or other assembly constructed of one or more component materials that are not defined as noncombustible.

[F] COMBUSTIBLE LIQUID. A liquid having a closed cup flash point at or above 100°F (38°C). Combustible liquids shall be subdivided as follows:

> **Class II.** Liquids having a closed cup flash point at or above 100°F (38°C) and below 140°F (60°C).

> **Class IIIA.** Liquids having a closed cup flash point at or above 140°F (60°C) and below 200°F (93°C).

> **Class IIIB.** Liquids having a closed cup flash point at or above 200°F (93°C).

The category of combustible liquids does not include compressed gases or cryogenic fluids.

COMBUSTIBLE MATERIAL. Any material not defined as noncombustible.

COMBUSTION. In the context of this code, refers to the rapid oxidation of fuel accompanied by the production of heat or heat and light.

COMBUSTION AIR. Air necessary for complete *combustion* of a fuel, including *theoretical air* and excess air.

COMBUSTION CHAMBER. The portion of an *appliance* within which *combustion* occurs.

COMBUSTION PRODUCTS. Constituents resulting from the *combustion* of a fuel with the oxygen of the air, including the inert gases, but excluding excess air.

COMMERCIAL COOKING APPLIANCES. Appliances used in a commercial food service establishment for heating or cooking food and which produce grease vapors, steam, fumes, smoke or odors that are required to be removed through a local exhaust ventilation system. Such appliances include deep fat fryers; upright broilers; griddles; broilers; steam-jacketed kettles; hot-top ranges; under-fired broilers (charbroilers); ovens; barbecues; rotisseries; and similar appliances. For the purpose of this definition, a food service establishment shall include any building or a portion thereof used for the preparation and serving of food.

COMMERCIAL COOKING RECIRCULATING SYSTEM. Self-contained system consisting of the exhaust hood, the cooking *equipment*, the filters and the fire suppression system. The system is designed to capture cooking vapors and residues generated from commercial cooking *equipment*. The system removes contaminants from the *exhaust air* and recirculates the air to the space from which it was withdrawn.

COMMERCIAL KITCHEN HOODS.

Backshelf hood. A backshelf hood is also referred to as a low-proximity hood, or as a sidewall hood where wall mounted. Its front lower lip is low over the *appliance*(s) and is "set back" from the front of the appliance(s). It is always closed to the rear of the appliances by a panel where free-standing, or by a panel or wall where wall mounted, and its height above the cooking surface varies. (This style of hood can be constructed with partial end panels to increase its effectiveness in capturing the effluent generated by the cooking operation).

Double island canopy hood. A double island canopy hood is placed over back-to-back appliances or *appliance* lines. It is open on all sides and overhangs both fronts and the sides of the *appliance*(s). It could have a wall panel between the backs of the appliances. (The fact that *exhaust air* is drawn from both sides of the double canopy to meet in the center causes each side of this hood to emulate a wall canopy hood, and thus it functions much the same with or without an actual wall panel between the backs of the appliances).

Eyebrow hood. An eyebrow hood is mounted directly to the face of an *appliance*, such as an oven and dishwasher, above the opening(s) or door(s) from which effluent is emitted, extending past the sides and overhanging the front of the opening to capture the effluent.

Pass-over hood. A pass-over hood is a free-standing form of a backshelf hood constructed low enough to pass food over the top.

Single island canopy hood. A single island canopy hood is placed over a single *appliance* or *appliance* line. It is open on all sides and overhangs the front, rear and sides of the *appliance*(s). A single island canopy is more susceptible to cross drafts and requires a greater *exhaust air* flow than an equivalent sized wall-mounted canopy to capture and contain effluent generated by the cooking operation(s).

Wall canopy hood. A wall canopy exhaust hood is mounted against a wall above a single *appliance* or line of *appliance*(s), or it could be free-standing with a back panel from the rear of the appliances to the hood. It overhangs the front and sides of the *appliance*(s) on all open sides.

The wall acts as a back panel, forcing the *makeup air* to be drawn across the front of the cooking *equipment*, thus increasing the effectiveness of the hood to capture and contain effluent generated by the cooking operation(s).

COMPENSATING HOODS. *Compensating hoods* are those having integral (built-in) *makeup air* supply. The *makeup air* supply for such hoods is generally supplied from: short-circuit flow from inside the hood, air curtain flow from the bottom of the front face, and front face discharge from the outside front wall of the hood. The compensating makeup airflow can also be supplied from the rear or side of the hood, or the rear, front or sides of the cooking *equipment*. The makeup airflow can be one or a combination of methods.

COMPRESSOR. A specific machine, with or without accessories, for compressing a gas.

COMPRESSOR, POSITIVE DISPLACEMENT. A compressor in which increase in pressure is attained by changing the internal volume of the compression chamber.

COMPRESSOR UNIT. A compressor with its prime mover and accessories.

CONCEALED LOCATION. A location that cannot be accessed without damaging permanent parts of the building structure or finish surface. Spaces above, below or behind readily removable panels or doors shall not be considered as concealed.

CONDENSATE. The liquid that condenses from a gas (including flue gas) caused by a reduction in temperature.

CONDENSER. A heat exchanger designed to liquefy refrigerant vapor by removal of heat.

CONDENSING UNIT. A specific refrigerating machine combination for a given refrigerant, consisting of one or more power-driven compressors, condensers and, where required, liquid receivers, and the regularly furnished accessories.

CONDITIONED SPACE. An area, room or space that is enclosed within the building thermal envelope and that is directly heated or cooled or that is indirectly heated or cooled. Spaces are indirectly heated or cooled where they communicate through openings with conditioned spaces, where they are separated from conditioned spaces by uninsulated walls, floors or ceilings, or where they contain uninsulated ducts, piping or other sources of heating or cooling.

[A] CONSTRUCTION DOCUMENTS. The written, graphic and pictorial documents prepared or assembled for describing the design, location and physical characteristics of the elements of the project necessary for obtaining a building permit. The construction drawings shall be drawn to an appropriate scale.

CONTROL. A manual or automatic device designed to regulate the gas, air, water or electrical supply to, or operation of, a mechanical system.

CONVERSION BURNER. A burner designed to supply gaseous fuel to an *appliance* originally designed to utilize another fuel.

COOKING APPLIANCE. See "*Commercial cooking appliances.*"

DAMPER. A manually or automatically controlled device to regulate draft or the rate of flow of air or *combustion* gases.

    Volume damper. A device that, where installed, will restrict, retard or direct the flow of air in a duct, or the products of *combustion* in a heat-producing *appliance*, its vent connector, vent or *chimney* therefrom.

[BS] DESIGN FLOOD ELEVATION. The elevation of the "design flood," including wave height, relative to the datum specified on the community's legally designated flood hazard area map. In areas designated as Zone AO, the *design flood elevation* shall be the elevation of the highest existing grade of the building's perimeter plus the depth number, in feet, specified on the flood hazard map. In areas designated as Zone AO where a depth number is not specified on the map, the depth number shall be taken as being equal to 2 feet (610 mm).

DESIGN WORKING PRESSURE. The maximum allowable working pressure for which a specific part of a system is designed.

DIRECT REFRIGERATION SYSTEM. A system in which the evaporator or condenser of the refrigerating system is in direct contact with the air or other substances to be cooled or heated.

DIRECT-VENT APPLIANCES. Appliances that are constructed and installed so that all air for *combustion* is derived from the outdoor atmosphere and all flue gases are discharged to the outdoor atmosphere.

DISCRETE PRODUCT. Products that are noncontinuous, individual, distinct pieces such as, but not limited to, electrical, plumbing and mechanical products and duct straps, duct fittings, duct registers and pipe hangers.

DRAFT. The pressure difference existing between the *appliance* or any component part and the atmosphere, that causes a continuous flow of air and products of *combustion* through the gas passages of the *appliance* to the atmosphere.

    Induced draft. The pressure difference created by the action of a fan, blower or ejector, that is located between the *appliance* and the *chimney* or vent termination.

    Natural draft. The pressure difference created by a vent or *chimney* because of its height, and the temperature difference between the flue gases and the atmosphere.

DRIP. The container placed at a low point in a system of piping to collect condensate and from which the condensate is removable.

DRY CLEANING SYSTEMS. Dry cleaning plants or systems are classified as follows:

    Type I. Those systems using Class I flammable liquid solvents having a flash point below 100°F (38°C).

    Type II. Those systems using Class II combustible liquid solvents having a flash point at or above 100°F (38°C) and below 140°F (60°C).

    Type III. Those systems using Class III combustible liquid solvents having a flash point at or above 140°F (60°C).

    Types IV and V. Those systems using Class IV nonflammable liquid solvents.

DUCT. A tube or conduit utilized for conveying air. The air passages of self-contained systems are not to be construed as air ducts.

DUCT FURNACE. A warm-air furnace normally installed in an air distribution duct to supply warm air for heating. This definition shall apply only to a warm-air heating *appliance* that, for air circulation, depends on a blower not furnished as part of the furnace.

DUCT SYSTEM. A continuous passageway for the transmission of air that, in addition to ducts, includes duct fittings, dampers, plenums, fans and accessory air-handling *equipment* and appliances.

DUCTLESS MINI-SPLIT SYSTEM. A heating and cooling system that is comprised of one or multiple indoor evaporator/air-handling units and an outdoor condensing unit that is connected by refrigerant piping and electrical wiring. A ductless mini-split system is capable of cooling or heating one or more rooms without the use of a traditional ductwork system.

[BG] DWELLING. A building or portion thereof that contains not more than two *dwelling* units.

[BG] DWELLING UNIT. A single unit providing complete, independent living facilities for one or more persons, including permanent provisions for living, sleeping, eating, cooking and sanitation.

ELECTRIC HEATING APPLIANCE. An *appliance* that produces heat energy to create a warm environment by the application of electric power to resistance elements, refrigerant compressors or dissimilar material junctions.

ENERGY RECOVERY VENTILATION SYSTEM. Systems that employ air-to-air heat exchangers to recover energy from or reject energy to *exhaust air* for the purpose of preheating, pre-cooling, humidifying or dehumidifying outdoor *ventilation air* prior to supplying such air to a space, either directly or as part of an HVAC system.

ENVIRONMENTAL AIR. Air that is conveyed to or from occupied areas through ducts which are not part of the heating or air-conditioning system, such as ventilation for human usage, domestic kitchen range exhaust, bathroom exhaust, domestic clothes dryer exhaust and parking garage exhaust.

EQUIPMENT. All piping, ducts, vents, control devices and other components of systems other than appliances which are permanently installed and integrated to provide control of environmental conditions for buildings. This definition shall also include other systems specifically regulated in this code.

EQUIPMENT, EXISTING. Any *equipment* regulated by this code which was legally installed prior to the effective date of this code, or for which a permit to install has been issued.

EVAPORATIVE COOLER. A device used for reducing the sensible heat of air for cooling by the process of evaporation of water into an airstream.

EVAPORATIVE COOLING SYSTEM. The *equipment* and appliances intended or installed for the purpose of environmental cooling by an evaporative cooler from which the conditioned air is distributed through ducts or plenums to the conditioned area.

EVAPORATOR. That part of the system in which liquid refrigerant is vaporized to produce refrigeration.

EXCESS AIR. The amount of air provided in addition to *theoretical air* to achieve complete *combustion* of a fuel, thereby preventing the formation of dangerous products of *combustion*.

EXFILTRATION. Uncontrolled outward air leakage from conditioned spaces through unintentional openings in ceilings, floors and walls to unconditioned spaces or the outdoors caused by pressure differences across these openings resulting from wind, the stack effect created by temperature differences between indoors and outdoors, and imbalances between supply and exhaust airflow rates.

EXHAUST SYSTEM. An assembly of connected ducts, plenums, fittings, registers, grilles and hoods through which air is conducted from the space or spaces and exhausted to the outdoor atmosphere.

EXTRA-HEAVY-DUTY COOKING APPLIANCE. Extra-heavy-duty cooking appliances are those utilizing open flame combustion of solid fuel at any time.

[BF] FIRE DAMPER. A *listed* device installed in ducts and air transfer openings designed to close automatically upon detection of heat and to restrict the passage of flame. Fire dampers are classified for use in either static systems that will automatically shut down in the event of a fire, or in dynamic systems that continue to operate during a fire. A dynamic fire damper is tested and rated for closure under elevated temperature airflow.

FIREPLACE. An assembly consisting of a hearth and fire chamber of noncombustible material and provided with a *chimney*, for use with solid fuels.

> **Factory-built fireplace.** A *listed* and *labeled* fireplace and *chimney* system composed of factory-made components, and assembled in the field in accordance with manufacturer's instructions and the conditions of the listing.

> **Masonry fireplace.** A field-constructed fireplace composed of solid masonry units, bricks, stones or concrete.

FIREPLACE STOVE. A free-standing chimney-connected solid-fuel-burning heater, designed to be operated with the fire chamber doors in either the open or closed position.

[FG] FLAME SAFEGUARD. A device that will automatically shut off the fuel supply to a main burner or group of burners when the means of ignition of such burners becomes inoperative, and when flame failure occurs on the burner or group of burners.

[BF] FLAME SPREAD INDEX. The numerical value assigned to a material tested in accordance with ASTM E 84 or UL 723.

FLAMMABILITY CLASSIFICATION. Refrigerants shall be assigned to one of the three classes—1, 2 or 3—in accordance with ASHRAE 34. For Classes 2 and 3, the heat of *combustion* shall be calculated assuming that *combustion* products are in the gas phase and in their most stable state.

> **Class 1.** Refrigerants that do not show flame propagation when tested in air at 14.7 psia (101 kPa) and 140°F (60°C).

> **Class 2.** Refrigerants having a lower flammability limit (LFL) of more than 0.00625 pound per cubic foot (0.10 kg/m3) at 140°F (60°C) and 14.7 psia (101 kPa) and a heat of combustion of less than 8169 Btu/lb (19 000 kJ/kg).

> **Class 3.** Refrigerants that are highly flammable, having a LFL of less than or equal to 0.00625 pound per cubic foot (0.10 kg/m3) at140°F (60°C) and 14.7 psia (101 kPa) or a heat of combustion greater than or equal to 8169 Btu/lb (19 000 kJ/kg).

[F] FLAMMABLE LIQUIDS. Any liquid that has a flash point below 100°F (38°C), and has a vapor pressure not exceeding 40 psia (276 kPa) at 100°F (38°C). Flammable liquids shall be known as Class I liquids and shall be divided into the following classifications:

> **Class IA.** Liquids having a flash point below 73°F (23°C) and a boiling point below 100°F (38°C).

> **Class IB.** Liquids having a flash point below 73°F (23°C) and a boiling point at or above 100°F (38°C).

> **Class IC.** Liquids having a flash point at or above 73°F (23°C) and below 100°F (38°C).

[F] FLAMMABLE VAPOR OR FUMES. Mixtures of gases in air at concentrations equal to or greater than the LFL and less than or equal to the upper flammability limit (UFL).

[F] FLASH POINT. The minimum temperature corrected to a pressure of 14.7 psia (101 kPa) at which the application of a test flame causes the vapors of a portion of the sample to ignite under the conditions specified by the test procedures and apparatus. The flash point of a liquid shall be determined in accordance with ASTM D 56, ASTM D 93 or ASTM D 3278.

FLEXIBLE AIR CONNECTOR. A conduit for transferring air between an air duct or plenum and an air terminal unit or between an air duct or plenum and an air inlet or air outlet. Such conduit is limited in its use, length and location.

FLOOR AREA, NET. The actual occupied area, not including unoccupied accessory areas or thicknesses of walls.

[FG] FLOOR FURNACE. A completely self-contained furnace suspended from the floor of the space being heated, taking air for *combustion* from outside such space and with means for observing flames and lighting the *appliance* from such space.

FLUE. A passageway within a *chimney* or vent through which gaseous *combustion* products pass.

FLUE CONNECTION (BREECHING). A passage for conducting the products of *combustion* from a fuel-fired *appliance* to the vent or *chimney* (see also "*Chimney* connector" and "Vent connector").

FLUE GASES. Products of *combustion* and excess air.

FLUE LINER (LINING). A system or material used to form the inside surface of a flue in a *chimney* or vent, for the purpose of protecting the surrounding structure from the effects of *combustion* products and conveying *combustion* products without leakage to the atmosphere.

[FG] FUEL GAS. A natural gas, manufactured gas, liquefied petroleum gas or a mixture of these.

FUEL OIL. Kerosene or any hydrocarbon oil having a flash point not less than 100°F (38°C).

FUEL-OIL PIPING SYSTEM. A closed piping system that connects a combustible liquid from a source of supply to a fuel-oil-burning *appliance*.

FURNACE. A completely self-contained heating unit that is designed to supply heated air to spaces remote from or adjacent to the *appliance* location.

FURNACE ROOM. A room primarily utilized for the installation of fuel-burning, space-heating and water-heating appliances other than boilers (see also "Boiler room").

FUSIBLE PLUG. A device arranged to relieve pressure by operation of a fusible member at a predetermined temperature.

GROUND SOURCE HEAT PUMP LOOP SYSTEM. Piping buried in horizontal or vertical excavations or placed in a body of water for the purpose of transporting heat transfer liquid to and from a heat pump. Included in this definition are closed loop systems in which the liquid is recirculated and open loop systems in which the liquid is drawn from a well or other source.

HAZARDOUS LOCATION. Any location considered to be a fire hazard for flammable vapors, dust, combustible fibers or other highly combustible substances. The location is not necessarily categorized in the *International Building Code* as a high-hazard use group classification.

HEAT EXCHANGER. A device that transfers heat from one medium to another.

HEAT PUMP. A refrigeration system that extracts heat from one substance and transfers it to another portion of the same substance or to a second substance at a higher temperature for a beneficial purpose.

HEAT TRANSFER LIQUID. The operating or thermal storage liquid in a mechanical system, including water or other liquid base, and additives at the concentration present under operating conditions used to move heat from one location to another. Refrigerants are not included as heat transfer liquids.

HEAVY-DUTY COOKING APPLIANCE. Heavy-duty cooking *appliances* include electric under-fired broilers, electric chain (conveyor) broilers, gas under-fired broilers, gas chain (conveyor) broilers, gas open-burner ranges (with or without oven), electric and gas wok ranges, smokers, smoker ovens, and electric and gas over-fired (upright) broilers and salamanders.

HIGH-PROBABILITY SYSTEMS. A refrigeration system in which the basic design or the location of components is such that a leakage of refrigerant from a failed connection, seal or component will enter an *occupancy* classified area, other than the *machinery room*.

HIGH-SIDE PRESSURE. The parts of a refrigerating system subject to condenser pressure.

HOOD. An air intake device used to capture by entrapment, impingement, adhesion or similar means, grease, moisture, heat and similar contaminants before they enter a duct system.

Type I. A kitchen hood for collecting and removing grease vapors and smoke. Such hoods are equipped with a fire suppression system.

Type II. A general kitchen hood for collecting and removing steam, vapor, heat, odors and products of *combustion*.

[FG] HYDROGEN GENERATING APPLIANCE. A self-contained package or factory-matched packages of integrated systems for generating gaseous hydrogen. Hydrogen generating appliances utilize electrolysis, reformation, chemical, or other processes to generate hydrogen.

IGNITION SOURCE. A flame, spark or hot surface capable of igniting flammable vapors or fumes. Such sources include *appliance* burners, burner ignitors and electrical switching devices.

[F] IMMEDIATELY DANGEROUS TO LIFE OR HEALTH (IDLH). The concentration of airborne contaminants that poses a threat of death, immediate or delayed permanent adverse health effects, or effects that could prevent escape from such an environment. This contaminant concentration level is established by the National Institute of Occupational Safety and Health (NIOSH) based on both toxicity and flammability. It is generally expressed in parts per million by volume (ppm v/v) or milligrams per cubic meter (mg/m³).

INDIRECT REFRIGERATION SYSTEM. A system in which a secondary coolant cooled or heated by the refrigerating system is circulated to the air or other substance to be cooled or heated. Indirect systems are distinguished by the following methods of application:

Closed system. A system in which a secondary fluid is either cooled or heated by the refrigerating system and then circulated within a closed circuit in indirect contact with the air or other substance to be cooled or heated.

Double-indirect open-spray system. A system in which the secondary substance for an indirect open-spray system is heated or cooled by an intermediate coolant circulated from a second enclosure.

Open-spray system. A system in which a secondary coolant is cooled or heated by the refrigerating system and then circulated in direct contact with the air or other substance to be cooled or heated.

Vented closed system. A system in which a secondary coolant is cooled or heated by the refrigerating system and then passed through a closed circuit in the air or other substance to be cooled or heated, except that the evaporator or condenser is placed in an open or appropriately vented tank.

INFILTRATION. Uncontrolled inward air leakage to conditioned spaces through unintentional openings in ceilings, floors and walls from unconditioned spaces or the outdoors caused by pressure differences across these openings resulting from wind, the stack effect created by temperature differences between indoors and outdoors, and imbalances between supply and exhaust airflow rates.

INTERLOCK. A device actuated by another device with which it is directly associated, to govern succeeding operations of the same or allied devices. A circuit in which a given action cannot occur until after one or more other actions have taken place.

JOINT, FLANGED. A joint made by bolting together a pair of flanged ends.

JOINT, FLARED. A metal-to-metal compression joint in which a conical spread is made on the end of a tube that is compressed by a flare nut against a mating flare.

JOINT, PLASTIC ADHESIVE. A joint made in thermoset plastic piping by the use of an adhesive substance which forms a continuous bond between the mating surfaces without dissolving either one of them.

JOINT, PLASTIC HEAT FUSION. A joint made in thermoplastic piping by heating the parts sufficiently to permit fusion of the materials when the parts are pressed together.

JOINT, PLASTIC SOLVENT CEMENT. A joint made in thermoplastic piping by the use of a solvent or solvent cement which forms a continuous bond between the mating surfaces.

JOINT, SOLDERED. A gas-tight joint obtained by the joining of metal parts with metallic mixtures of alloys which melt at temperatures between 400°F (204°C) and 1,000°F (538°C).

JOINT, WELDED. A gas-tight joint obtained by the joining of metal parts in molten state.

[A] LABELED. *Equipment*, materials or products to which have been affixed a label, seal, symbol or other identifying mark of a nationally recognized testing laboratory, inspection agency or other organization concerned with product evaluation that maintains periodic inspection of the production of the above-labeled items and whose labeling indicates either that the *equipment*, material or product meets identified standards or has been tested and found suitable for a specified purpose.

LIGHT-DUTY COOKING APPLIANCE. Light-duty cooking *appliances* include gas and electric ovens (including standard, bake, roasting, revolving, retherm, convection, combination convection/steamer, countertop conveyorized baking/finishing, deck and pastry), electric and gas steam-jacketed kettles, electric and gas pasta cookers, electric and gas compartment steamers (both pressure and atmospheric) and electric and gas cheesemelters.

[FG] LIMIT CONTROL. A device responsive to changes in pressure, temperature or level for turning on, shutting off or throttling the gas supply to an *appliance*.

LIMITED CHARGE SYSTEM. A system in which, with the compressor idle, the design pressure will not be exceeded when the refrigerant charge has completely evaporated.

[A] LISTED. *Equipment*, materials, products or services included in a list published by an organization acceptable to the code official and concerned with evaluation of products or services that maintains periodic inspection of production of *listed equipment* or materials or periodic evaluation of services and whose listing states either that the *equipment*, mate-

rial, product or service meets identified standards or has been tested and found suitable for a specified purpose.

LIVING SPACE. Space within a *dwelling unit* utilized for living, sleeping, eating, cooking, bathing, washing and sanitation purposes.

LOWER EXPLOSIVE LIMIT (LEL). See "LFL."

LOWER FLAMMABLE LIMIT (REFRIGERANT) (LFL). The minimum concentration of refrigerant that is capable of propagating a flame through a homogeneous mixture of refrigerant and air.

[F] LOWER FLAMMABLE LIMIT (LFL). The minimum concentration of vapor in air at which propagation of flame will occur in the presence of an ignition source. The LFL is sometimes referred to as LEL or lower explosive limit.

LOW-PRESSURE HOT-WATER-HEATING BOILER. A boiler furnishing hot water at pressures not exceeding 160 psi (1103 kPa) and at temperatures not exceeding 250°F (121°C).

LOW-PRESSURE STEAM-HEATING BOILER. A boiler furnishing steam at pressures not exceeding 15 psi (103 kPa).

LOW-PROBABILITY SYSTEMS. A refrigeration system in which the basic design or the location of components is such that a leakage of refrigerant from a failed connection, seal or component will not enter an occupancy-classified area, other than the *machinery room*.

LOW-SIDE PRESSURE. The parts of a refrigerating system subject to evaporator pressure.

MACHINERY ROOM. A room meeting prescribed safety requirements and in which refrigeration systems or components thereof are located (see Sections 1105 and 1106).

MECHANICAL DRAFT SYSTEM. A venting system designed to remove flue or vent gases by mechanical means, that consists of an induced-draft portion under nonpositive static pressure or a forced-draft portion under positive static pressure.

> **Forced-draft venting system.** A portion of a venting system using a fan or other mechanical means to cause the removal of flue or vent gases under positive static pressure.

> **Induced-draft venting system.** A portion of a venting system using a fan or other mechanical means to cause the removal of flue or vent gases under nonpositive static vent pressure.

> **Power venting system.** A portion of a venting system using a fan or other mechanical means to cause the removal of flue or vent gases under positive static vent pressure.

MECHANICAL EQUIPMENT/APPLIANCE ROOM. A room or space in which nonfuel-fired mechanical *equipment* and *appliances* are located.

MECHANICAL EXHAUST SYSTEM. A system for removing air from a room or space by mechanical means.

MECHANICAL JOINT.

1. A connection between pipes, fittings, or pipes and fittings that is not welded, brazed, caulked, soldered, solvent cemented or heat fused.

2. A general form of gas or liquid-tight connections obtained by the joining of parts through a positive holding mechanical construction such as, but not limited to, flanged, screwed, clamped or flared connections.

MECHANICAL SYSTEM. A system specifically addressed and regulated in this code and composed of components, devices, *appliances* and *equipment*.

MEDIUM-DUTY COOKING APPLIANCE. Medium-duty cooking *appliances* include electric discrete element ranges (with or without oven), electric and gas hot-top ranges, electric and gas griddles, electric and gas double-sided griddles, electric and gas fryers (including open deep fat fryers, donut fryers, kettle fryers and pressure fryers), electric and gas conveyor pizza ovens, electric and gas tilting skillets (braising pans) and electric and gas rotisseries.

MODULAR BOILER. A steam or hot-water-heating assembly consisting of a group of individual boilers called modules intended to be installed as a unit without intervening stop valves. Modules are under one jacket or are individually jacketed. The individual modules shall be limited to a maximum input rating of 400,000 Btu/h (117 228 W) gas, 3 gallons per hour (gph) (11.4 L/h) oil, or 115 kW (electric).

NATURAL DRAFT SYSTEM. A venting system designed to remove flue or vent gases under nonpositive static vent pressure entirely by natural draft.

NATURAL VENTILATION. The movement of air into and out of a space through intentionally provided openings, such as windows and doors, or through nonpowered ventilators.

NET OCCUPIABLE FLOOR AREA. The floor area of an *occupiable space* defined by the inside surfaces of its walls but excluding shafts, column enclosures and other permanently enclosed, inaccessible and unoccupiable areas. Obstructions in the space such as furnishings, display or storage racks and other obstructions, whether temporary or permanent, shall not be deducted from the space area.

NONABRASIVE/ABRASIVE MATERIALS. Nonabrasive particulate in high concentrations, moderately abrasive particulate in low and moderate concentrations, and highly abrasive particulate in low concentrations, such as alfalfa, asphalt, plaster, gypsum and salt.

NONCOMBUSTIBLE MATERIALS. Materials that, when tested in accordance with ASTM E 136, have at least three of four specimens tested meeting all of the following criteria:

1. The recorded temperature of the surface and interior thermocouples shall not at any time during the test rise more than 54°F (30°C) above the furnace temperature at the beginning of the test.

2. There shall not be flaming from the specimen after the first 30 seconds.

3. If the weight loss of the specimen during testing exceeds 50 percent, the recorded temperature of the surface and interior thermocouples shall not at any time during the test rise above the furnace air temperature at the beginning of the test, and there shall not be flaming of the specimen.

[A] OCCUPANCY. The purpose for which a building, or portion thereof, is utilized or occupied.

OCCUPATIONAL EXPOSURE LIMIT (OEL). The time-weighted average (TWA) concentration for a normal eight-hour workday and a 40-hour workweek to which nearly all workers can be repeatedly exposed without adverse effect, based on the OSHA PEL, ACGIH TLV-TWA, AIHA WEEL, or consistent value.

OCCUPIABLE SPACE. An enclosed space intended for human activities, excluding those spaces intended primarily for other purposes, such as storage rooms and *equipment* rooms, that are only intended to be occupied occasionally and for short periods of time.

OFFSET (VENT). A combination of *approved* bends that make two changes in direction bringing one section of the vent out of line but into a line parallel with the other section.

OUTDOOR AIR. Air taken from the outdoors, and therefore not previously circulated through the system.

OUTDOOR OPENING. A door, window, louver or skylight openable to the outdoor atmosphere.

OUTLET. A threaded connection or bolted flange in a piping system to which a gas-burning *appliance* is attached.

PANEL HEATING. A method of radiant space heating in which heat is supplied by large heated areas of room surfaces. The heating element usually consists of warm water piping, warm air ducts, or electrical resistance elements embedded in or located behind ceiling, wall or floor surfaces.

PELLET FUEL-BURNING APPLIANCE. A closed-combustion, vented *appliance* equipped with a fuel-feed mechanism for burning processed pellets of solid fuel of a specified size and composition.

PIPING. Where used in this code, "piping" refers to either pipe or tubing, or both.

> **Pipe.** A rigid conduit of iron, steel, copper, brass or plastic.

> **Tubing.** Semirigid conduit of copper, aluminum, plastic or steel.

PLASTIC, THERMOPLASTIC. A plastic that is capable of being repeatedly softened by increase of temperature and hardened by decrease of temperature.

PLASTIC, THERMOSETTING. A plastic that is capable of being changed into a substantially infusible or insoluble product when cured under application of heat or chemical means.

PLENUM. An enclosed portion of the building structure, other than an *occupiable space* being conditioned, that is designed to allow air movement, and thereby serve as part of an air distribution system.

PORTABLE FUEL CELL APPLIANCE. A fuel cell generator of electricity, which is not fixed in place. A portable fuel cell *appliance* utilizes a cord and plug connection to a grid-isolated load and has an integral fuel supply.

POWER BOILER. See "Boiler."

[A] PREMISES. A lot, plot or parcel of land, including any structure thereon.

PRESS JOINT. A permanent mechanical joint incorporating an elastomeric seal or an elastomeric seal and corrosion-resistant grip ring. The joint is made with a pressing tool and jaw or ring approved by the fitting manufacturer.

PRESSURE, FIELD TEST. A test performed in the field to prove system tightness.

PRESSURE-LIMITING DEVICE. A pressure-responsive mechanism designed to stop automatically the operation of the pressure-imposing element at a predetermined pressure.

PRESSURE RELIEF DEVICE. A pressure-actuated valve or rupture member designed to relieve excessive pressure automatically.

PRESSURE RELIEF VALVE. A pressure-actuated valve held closed by a spring or other means and designed to relieve pressure automatically in excess of the device's setting.

PRESSURE VESSELS. Closed containers, tanks or vessels that are designed to contain liquids or gases, or both, under pressure.

PRESSURE VESSELS—REFRIGERANT. Any refrigerant-containing receptacle in a refrigerating system. This does not include evaporators where each separate section does not exceed 0.5 cubic foot (0.014 m^3) of refrigerant-containing volume, regardless of the maximum inside dimensions, evaporator coils, controls, headers, pumps and piping.

PROTECTIVE ASSEMBLY (REDUCED CLEARANCE). Any noncombustible assembly that is *labeled* or constructed in accordance with Table 308.4.2 and is placed between combustible materials or assemblies and mechanical appliances, devices or *equipment*, for the purpose of reducing required airspace clearances. Protective assemblies attached directly to a combustible assembly shall not be considered as part of that combustible assembly.

PURGE. To clear of air, water or other foreign substances.

PUSH-FIT JOINTS. A type of mechanical joint consisting of elastomeric seals and corrosion-resistant tube grippers. Such joints are permanent or removable depending on the design.

QUICK-OPENING VALVE. A valve that opens completely by fast action, either manually or automatically controlled. A valve requiring one-quarter round turn or less is considered to be quick opening.

RADIANT HEATER. A heater designed to transfer heat primarily by direct radiation.

READY ACCESS (TO). That which enables a device, *appliance* or *equipment* to be directly reached, without requiring the removal or movement of any panel, door or similar obstruction [see "Access (to)"].

RECEIVER, LIQUID. A vessel permanently connected to a refrigeration system by inlet and outlet pipes for storage of liquid refrigerant.

RECIRCULATED AIR. Air removed from a conditioned space and intended for reuse as supply air.

RECLAIMED REFRIGERANTS. Refrigerants reprocessed to the same specifications as for new refrigerants by means including distillation. Such refrigerants have been chemically analyzed to verify that the specifications have been met. Reclaiming usually implies the use of processes or procedures that are available only at a reprocessing or manufacturing facility.

RECOVERED REFRIGERANTS. Refrigerants removed from a system in any condition without necessarily testing or processing them.

RECYCLED REFRIGERANTS. Refrigerants from which contaminants have been reduced by oil separation, removal of noncondensable gases, and single or multiple passes through devices that reduce moisture, acidity and particulate matter, such as replaceable core filter driers. These procedures usually are performed at the field job site or in a local service shop.

REFRIGERANT. A substance utilized to produce refrigeration by its expansion or vaporization.

REFRIGERANT SAFETY CLASSIFICATIONS. Groupings that indicate the toxicity and flammability classes in accordance with Section 1103.1. The classification group is made up of a letter (A or B) that indicates the toxicity class, followed by a number (1, 2 or 3) that indicates the flammability class. Refrigerant blends are similarly classified, based on the compositions at their worst cases of fractionation, as separately determined for toxicity and flammability. In some cases, the worst case of fractionation is the original formulation.

    Flammability. See "Flammability classification."

    Toxicity. See "Toxicity classification."

REFRIGERATED ROOM OR SPACE. A room or space in which an evaporator or brine coil is located for the purpose of reducing or controlling the temperature within the room or space to below 68°F (20°C).

REFRIGERATING SYSTEM. A combination of interconnected refrigerant-containing parts constituting one closed refrigerant circuit in which a refrigerant is circulated for the purpose of extracting heat.

REFRIGERATION CAPACITY RATING. Expressed as 1 horsepower (0.75 kW), 1 ton or 12,000 Btu/h (3.5 kW), shall all mean the same quantity.

REFRIGERATION MACHINERY ROOM. See *Machinery room.*

REFRIGERATION SYSTEM, ABSORPTION. A heat-operated, closed-refrigeration cycle in which a secondary fluid (the absorbent) absorbs a primary fluid (the refrigerant) that has been vaporized in the evaporator.

    Direct system. A system in which the evaporator is in direct contact with the material or space refrigerated, or is located in air-circulating passages communicating with such spaces.

Indirect system. A system in which a brine coil cooled by the refrigerant is circulated to the material or space refrigerated, or is utilized to cool the air so circulated. Indirect systems are distinguished by the type or method of application.

REFRIGERATION SYSTEM CLASSIFICATION. Refrigeration systems are classified according to the degree of probability that leaked refrigerant from a failed connection, seal or component will enter an occupied area. The distinction is based on the basic design or location of the components.

REFRIGERATION SYSTEM, MECHANICAL. A combination of interconnected refrigeration-containing parts constituting one closed refrigerant circuit in which a refrigerant is circulated for the purpose of extracting heat and in which a compressor is used for compressing the refrigerant vapor.

REFRIGERATION SYSTEM, SELF-CONTAINED. A complete factory-assembled and tested system that is shipped in one or more sections and that does not have refrigerant-containing parts that are joined in the field by other than companion or block valves.

[A] REGISTERED DESIGN PROFESSIONAL. An individual who is registered or licensed to practice their respective design profession as defined by the statutory requirements of the professional registration laws of the state or jurisdiction in which the project is to be constructed.

RETURN AIR. Air removed from an *approved* conditioned space or location and recirculated or exhausted.

RETURN AIR SYSTEM. An assembly of connected ducts, plenums, fittings, registers and grilles through which air from the space or spaces to be heated or cooled is conducted back to the supply unit (see also "Supply air system").

[FG] ROOM HEATER VENTED. A free-standing heating unit burning solid or liquid fuel for direct heating of the space in and adjacent to that in which the unit is located.

SAFETY VALVE. A valve that relieves pressure in a steam boiler by opening fully at the rated discharge pressure. The valve is of the spring-pop type.

SELF-CONTAINED EQUIPMENT. Complete, factory-assembled and tested, heating, air-conditioning or refrigeration *equipment* installed as a single unit, and having all working parts, complete with motive power, in an enclosed unit of said machinery.

[BF] SHAFT. An enclosed space extending through one or more stories of a building, connecting vertical openings in successive floors, or floors and the roof.

[BF] SHAFT ENCLOSURE. The walls or construction forming the boundaries of a shaft.

[BG] SLEEPING UNIT. A room or space in which people sleep, which can also include permanent provisions for living, eating, and either sanitation or kitchen facilities but not both. Such rooms and spaces that are also part of a *dwelling unit* are not sleeping units.

[BF] SMOKE DAMPER. A *listed* device installed in ducts and air transfer openings designed to resist the passage of smoke. The device is installed to operate automatically, controlled by a smoke detection system, and where required, is capable of being positioned from a fire command center.

[BF] SMOKE-DEVELOPED INDEX. A numerical value assigned to a material tested in accordance with ASTM E 84.

SOLID FUEL (COOKING APPLICATIONS). Applicable to commercial food service operations only, solid fuel is any bulk material such as hardwood, mesquite, charcoal or briquettes that is combusted to produce heat for cooking operations.

SOURCE CAPTURE SYSTEM. A mechanical exhaust system designed and constructed to capture air contaminants at their source and to exhaust such contaminants to the outdoor atmosphere.

[FG] STATIONARY FUEL CELL POWER PLANT. A self-contained package or factory-matched packages which constitute an automatically operated assembly of integrated systems for generating useful electrical energy and recoverable thermal energy that is permanently connected and fixed in place.

STEAM-HEATING BOILER. A boiler operated at pressures not exceeding 15 psi (103 kPa) for steam.

STOP VALVE. A shutoff valve for controlling the flow of liquid or gases.

[BG] STORY. That portion of a building included between the upper surface of a floor and the upper surface of the floor next above, except that the topmost story shall be that portion of a building included between the upper surface of the topmost floor and the ceiling or roof above.

STRENGTH, ULTIMATE. The highest stress level that the component will tolerate without rupture.

SUPPLY AIR. That air delivered to each or any space supplied by the air distribution system or the total air delivered to all spaces supplied by the air distribution system, which is provided for ventilating, heating, cooling, humidification, dehumidification and other similar purposes.

SUPPLY AIR SYSTEM. An assembly of connected ducts, plenums, fittings, registers and grilles through which air, heated or cooled, is conducted from the supply unit to the space or spaces to be heated or cooled (see also "Return air system").

THEORETICAL AIR. The exact amount of air required to supply oxygen for complete *combustion* of a given quantity of a specific fuel.

THERMAL RESISTANCE (R). A measure of the ability to retard the flow of heat. The R-value is the reciprocal of thermal conductance.

[P] THIRD-PARTY CERTIFICATION AGENCY. An approved agency operating a product or material certification system that incorporates initial product testing, assessment and surveillance of a manufacturer's quality control system.

[P] THIRD-PARTY CERTIFIED. Certification obtained by the manufacturer indicating that the function and performance characteristics of a product or material have been determined by testing and ongoing surveillance by an approved third-party certification agency. Assertion of certification is in the form of identification in accordance with the requirements of the third-party certification agency.

[P] THIRD-PARTY TESTED. Procedure by which an approved testing laboratory provides documentation that a product, material or system conforms to specified requirements.

TLV-TWA (THRESHOLD LIMIT VALUE-TIME-WEIGHTED AVERAGE). The time-weighted average concentration of a refrigerant or other chemical in air for a normal 8-hour workday and a 40-hour workweek, to which nearly all workers are repeatedly exposed, day after day, without adverse effects, as adopted by the American Conference of Government Industrial Hygienists (ACGIH).

TOILET ROOM. A room containing a water closet and, frequently, a lavatory, but not a bathtub, shower, spa or similar bathing fixture.

TOXICITY CLASSIFICATION. Refrigerants shall be classified for toxicity in one of two classes in accordance with ASHRAE 34:

Class A. Refrigerants that have an occupational exposure limit (OEL) of 400 parts per million (ppm) or greater.

Class B. Refrigerants that have an OEL of less than 400 ppm.

TRANSITION FITTINGS, PLASTIC TO STEEL. An adapter for joining plastic pipe to steel pipe. The purpose of this fitting is to provide a permanent, pressure-tight connection between two materials which cannot be joined directly one to another.

UNIT HEATER. A self-contained *appliance* of the fan type, designed for the delivery of warm air directly into the space in which the *appliance* is located.

VENT. A pipe or other conduit composed of factory-made components, containing a passageway for conveying *combustion* products and air to the atmosphere, *listed* and *labeled* for use with a specific type or class of *appliance*.

Pellet vent. A vent *listed* and *labeled* for use with *listed* pellet-fuel-burning appliances.

Type L vent. A vent *listed* and *labeled* for use with the following:

1. Oil-burning appliances that are *listed* for use with Type L vents.

2. Gas-fired appliances that are *listed* for use with Type B vents.

VENT CONNECTOR. The pipe that connects an *approved* fuel-fired *appliance* to a vent.

VENT DAMPER DEVICE, AUTOMATIC. A device intended for installation in the venting system, in the outlet of an individual automatically operated fuel-burning *appliance* that is designed to open the venting system automatically when the *appliance* is in operation and to close off the venting system automatically when the *appliance* is in a standby or shutdown condition.

VENTILATION. The natural or mechanical process of supplying conditioned or unconditioned air to, or removing such air from, any space.

VENTILATION AIR. That portion of supply air that comes from the outside (outdoors), plus any recirculated air that has been treated to maintain the desired quality of air within a designated space.

VENTING SYSTEM. A continuous open passageway from the flue collar of an *appliance* to the outdoor atmosphere for the purpose of removing flue or vent gases. A venting system is usually composed of a vent or a *chimney* and vent connector, if used, assembled to form the open passageway.

WATER HEATER. Any heating *appliance* or *equipment* that heats potable water and supplies such water to the potable hot water distribution system.

ZONE. One *occupiable space* or several occupiable spaces with similar *occupancy* classification (see Table 403.3.1.1), occupant density, zone air distribution effectiveness and zone primary airflow rate per unit area.

CHAPTER 3

GENERAL REGULATIONS

SECTION 301
GENERAL

301.1 Scope. This chapter shall govern the approval and installation of all *equipment* and appliances that comprise parts of the building mechanical systems regulated by this code in accordance with Section 101.2.

301.2 Energy utilization. Heating, ventilating and air-conditioning systems of all structures shall be designed and installed for efficient utilization of energy in accordance with the *International Energy Conservation Code*.

301.3 Identification. Each length of pipe and tubing and each pipe fitting utilized in a mechanical system shall bear the identification of the manufacturer.

301.4 Plastic pipe, fittings and components. Plastic pipe, fittings and components shall be *third-party certified* as conforming to NSF 14.

301.5 Third-party testing and certification. Piping, tubing and fittings shall comply with the applicable referenced standards, specifications and performance criteria of this code and shall be identified in accordance with Section 301.3. Piping, tubing and fittings shall either be tested by an approved third-party testing agency or certified by an approved *third-party certification agency*.

301.6 Fuel gas appliances and equipment. The approval and installation of fuel gas distribution piping and *equipment*, fuel gas-fired appliances and fuel gas-fired *appliance* venting systems shall be in accordance with the *International Fuel Gas Code*.

301.7 Listed and labeled. Appliances regulated by this code shall be *listed* and *labeled* for the application in which they are installed and used, unless otherwise *approved* in accordance with Section 105.

> **Exception:** Listing and labeling of *equipment* and appliances used for refrigeration shall be in accordance with Section 1101.2.

301.8 Labeling. Labeling shall be in accordance with the procedures set forth in Sections 301.8.1 through 301.8.2.3.

301.8.1 Testing. An *approved* agency shall test a representative sample of the mechanical *equipment* and appliances being *labeled* to the relevant standard or standards. The *approved* agency shall maintain a record of all of the tests performed. The record shall provide sufficient detail to verify compliance with the test standard.

301.8.2 Inspection and identification. The *approved* agency shall periodically perform an inspection, which shall be in-plant if necessary, of the mechanical *equipment* and appliances to be *labeled*. The inspection shall verify that the *labeled* mechanical *equipment* and appliances are representative of the mechanical *equipment* and appliances tested.

301.8.2.1 Independent. The agency to be *approved* shall be objective and competent. To confirm its objectivity, the agency shall disclose all possible conflicts of interest.

301.8.2.2 Equipment. An *approved* agency shall have adequate *equipment* to perform all required tests. The *equipment* shall be periodically calibrated.

301.8.2.3 Personnel. An *approved* agency shall employ experienced personnel educated in conducting, supervising and evaluating tests.

301.9 Label information. A permanent factory-applied nameplate(s) shall be affixed to appliances on which shall appear in legible lettering, the manufacturer's name or trademark, the model number, serial number and the seal or mark of the *approved* agency. A label shall also include the following:

1. Electrical *equipment* and appliances: Electrical rating in volts, amperes and motor phase; identification of individual electrical components in volts, amperes or watts, motor phase; Btu/h (W) output; and required clearances.

2. Absorption units: Hourly rating in Btu/h (W); minimum hourly rating for units having step or automatic modulating controls; type of fuel; type of refrigerant; cooling capacity in Btu/h (W); and required clearances.

3. Fuel-burning units: Hourly rating in Btu/h (W); type of fuel *approved* for use with the *appliance*; and required clearances.

4. Electric comfort heating appliances: electric rating in volts, amperes and phase; Btu/h (W) output rating; individual marking for each electrical component in amperes or watts, volts and phase; and required clearances from combustibles.

301.10 Electrical. Electrical wiring, controls and connections to *equipment* and appliances regulated by this code shall be in accordance with NFPA 70.

301.11 Plumbing connections. Potable water supply and building drainage system connections to *equipment* and appliances regulated by this code shall be in accordance with the *International Plumbing Code*.

301.12 Fuel types. Fuel-fired appliances shall be designed for use with the type of fuel to which they will be connected and the altitude at which they are installed. Appliances that comprise parts of the building mechanical system shall not be converted for the usage of a different fuel, except where *approved* and converted in accordance with the manufacturer's instructions. The fuel input rate shall not be increased or decreased beyond the limit rating for the altitude at which the *appliance* is installed.

301.13 Vibration isolation. Where vibration isolation of *equipment* and appliances is employed, an *approved* means of supplemental restraint shall be used to accomplish the support and restraint.

301.14 Repair. Defective material or parts shall be replaced or repaired in such a manner so as to preserve the original approval or listing.

301.15 Wind resistance. Mechanical *equipment*, appliances and supports that are exposed to wind shall be designed and installed to resist the wind pressures determined in accordance with the *International Building Code*.

[BS] 301.16 Flood hazard. For structures located in flood hazard areas, mechanical systems, equipment and appliances shall be located at or above the elevation required by Section 1612 of the *International Building Code* for utilities and attendant equipment.

> **Exception:** Mechanical systems, equipment and appliances are permitted to be located below the elevation required by Section 1612 of the of the *International Building Code* for utilities and attendant equipment provided that they are designed and installed to prevent water from entering or accumulating within the components and to resist hydrostatic and hydrodynamic loads and stresses, including the effects of buoyancy, during the occurrence of flooding up to such elevation.

[BS] 301.16.1 Coastal high-hazard areas and coastal A zones. In coastal high-hazard areas and coastal A zones, mechanical systems and *equipment* shall not be mounted on or penetrate walls intended to break away under flood loads.

301.17 Rodentproofing. Buildings or structures and the walls enclosing habitable or occupiable rooms and spaces in which persons live, sleep or work, or in which feed, food or foodstuffs are stored, prepared, processed, served or sold, shall be constructed to protect against the entrance of rodents in accordance with the *International Building Code*.

301.18 Seismic resistance. Where earthquake loads are applicable in accordance with the *International Building Code*, mechanical system supports shall be designed and installed for the seismic forces in accordance with the *International Building Code*.

SECTION 302
PROTECTION OF STRUCTURE

302.1 Structural safety. The building or structure shall not be weakened by the installation of mechanical systems. Where floors, walls, ceilings or any other portion of the building or structure are required to be altered or replaced in the process of installing or repairing any system, the building or structure shall be left in a safe structural condition in accordance with the *International Building Code*.

302.2 Penetrations of floor/ceiling assemblies and fire-resistance-rated assemblies. Penetrations of floor/ceiling assemblies and assemblies required to have a fire-resistance rating shall be protected in accordance with Chapter 7 of the *International Building Code*.

[BS] 302.3 Cutting, notching and boring in wood framing. The cutting, notching and boring of wood framing members shall comply with Sections 302.3.1 through 302.3.4.

[BS] 302.3.1 Joist notching. Notches on the ends of joists shall not exceed one-fourth the joist depth. Holes bored in joists shall not be within 2 inches (51 mm) of the top or bottom of the joist, and the diameter of any such hole shall not exceed one-third the depth of the joist. Notches in the top or bottom of joists shall not exceed one-sixth the depth and shall not be located in the middle third of the span.

[BS] 302.3.2 Stud cutting and notching. In exterior walls and bearing partitions, a wood stud shall not be cut or notched in excess of 25 percent of its depth. In nonbearing partitions that do not support loads other than the weight of the partition, a stud shall not be cut or notched in excess of 40 percent of its depth.

[BS] 302.3.3 Bored holes. The diameter of bored holes in wood studs shall not exceed 40 percent of the stud depth. The diameter of bored holes in wood studs shall not exceed 60 percent of the stud depth in nonbearing partitions. The diameter of bored holes in wood studs shall not exceed 60 percent of the stud depth in any wall where each stud is doubled, provided that not more than two such successive doubled studs are so bored. The edge of the bored hole shall be not closer than $\frac{5}{8}$ inch (15.9 mm) to the edge of the stud. Bored holes shall be not located at the same section of stud as a cut or notch.

[BS] 302.3.4 Engineered wood products. Cuts, notches and holes bored in trusses, structural composite lumber, structural glue-laminated members and I-joists are prohibited except where permitted by the manufacturer's recommendations or where the effects of such alterations are specifically considered in the design of the member by a registered design professional.

[BS] 302.4 Alterations to trusses. Truss members and components shall not be cut, drilled, notched, spliced or otherwise altered in any way without written concurrence and approval of a *registered design professional*. Alterations resulting in the addition of loads to any member, such as HVAC *equipment* and water heaters, shall not be permitted without verification that the truss is capable of supporting such additional loading.

[BS] 302.5 Cutting, notching and boring in steel framing. The cutting, notching and boring of steel framing members shall comply with Sections 302.5.1 through 302.5.3.

[BS] 302.5.1 Cutting, notching and boring holes in structural steel framing. The cutting, notching and boring of holes in structural steel framing members shall be as prescribed by the *registered design professional*.

[BS] 302.5.2 Cutting, notching and boring holes in cold-formed steel framing. Flanges and lips of load-bearing cold-formed steel framing members shall not be cut or notched. Holes in webs of load-bearing cold-formed steel framing members shall be permitted along the centerline of the web of the framing member and shall not exceed the dimensional limitations, penetration spacing or minimum hole edge distance as prescribed by the *registered design professional*. Cutting, notching and boring holes of steel floor/roof decking shall be as prescribed by the *registered design professional*.

[BS] 302.5.3 Cutting, notching and boring holes in non-structural cold-formed steel wall framing. Flanges and lips of nonstructural cold-formed steel wall studs shall not be cut or notched. Holes in webs of nonstructural cold-formed steel wall studs shall be permitted along the center-line of the web of the framing member, shall not exceed $1^1/_2$ inches (38 mm) in width or 4 inches (102 mm) in length, and shall not be spaced less than 24 inches (610 mm) center to center from another hole or less than 10 inches (254 mm) from the bearing end.

SECTION 303
EQUIPMENT AND APPLIANCE LOCATION

303.1 General. *Equipment* and appliances shall be located as required by this section, specific requirements elsewhere in this code and the conditions of the *equipment* and *appliance* listing.

303.2 Hazardous locations. Appliances shall not be located in a *hazardous location* unless *listed* and *approved* for the specific installation.

303.3 Prohibited locations. Fuel-fired appliances shall not be located in, or obtain *combustion* air from, any of the following rooms or spaces:

1. Sleeping rooms.

2. Bathrooms.

3. Toilet rooms.

4. Storage closets.

5. Surgical rooms.

Exception: This section shall not apply to the following appliances:

1. *Direct-vent* appliances that obtain all *combustion* air directly from the outdoors.

2. Solid fuel-fired appliances, provided that combustion air is provided in accordance with the manufacturers' instructions.

3. Appliances installed in a dedicated enclosure in which all *combustion* air is taken directly from the outdoors, in accordance with Chapter 7. *Access* to such enclosure shall be through a solid door, weather-stripped in accordance with the exterior door air leakage requirements of the *International Energy Conservation Code* and equipped with an *approved* self-closing device.

303.4 Protection from damage. Appliances shall not be installed in a location where subject to mechanical damage unless protected by *approved* barriers.

303.5 Indoor locations. Furnaces and boilers installed in closets and alcoves shall be listed for such installation.

303.6 Outdoor locations. Appliances installed in other than indoor locations shall be *listed* and *labeled* for outdoor installation.

303.7 Pit locations. Appliances installed in pits or excavations shall not come in direct contact with the surrounding soil. The sides of the pit or excavation shall be held back not less than 12 inches (305 mm) from the *appliance*. Where the depth exceeds 12 inches (305 mm) below adjoining grade, the walls of the pit or excavation shall be lined with concrete or masonry. Such concrete or masonry shall extend not less than 4 inches (102 mm) above adjoining grade and shall have sufficient lateral load-bearing capacity to resist collapse. The *appliance* shall be protected from flooding in an *approved* manner.

[BF] 303.8 Elevator shafts. Mechanical systems shall not be located in an elevator shaft.

SECTION 304
INSTALLATION

304.1 General. *Equipment* and appliances shall be installed as required by the terms of their approval, in accordance with the conditions of the listing, the manufacturer's installation instructions and this code. Manufacturer's installation instructions shall be available on the job site at the time of inspection.

304.2 Conflicts. Where conflicts between this code and the conditions of listing or the manufacturer's installation instructions occur, the provisions of this code shall apply.

Exception: Where a code provision is less restrictive than the conditions of the listing of the *equipment* or *appliance* or the manufacturer's installation instructions, the conditions of the listing and the manufacturer's installation instructions shall apply.

304.3 Elevation of ignition source. Equipment and appliances having an *ignition source* and located in hazardous locations and public garages, private garages, repair garages, automotive motor fuel-dispensing facilities and parking garages shall be elevated such that the source of ignition is not less than 18 inches (457 mm) above the floor surface on which the *equipment* or *appliance* rests. For the purpose of this section, rooms or spaces that are not part of the living space of a *dwelling unit* and that communicate directly with a private garage through openings shall be considered to be part of the private garage.

Exception: Elevation of the ignition source is not required for appliances that are listed as flammable vapor ignition resistant.

304.3.1 Parking garages. Connection of a parking garage with any room in which there is a fuel-fired *appliance* shall be by means of a vestibule providing a two-doorway separation, except that a single door is permitted where the sources of ignition in the *appliance* are elevated in accordance with Section 304.3.

Exception: This section shall not apply to *appliance* installations complying with Section 304.6.

304.4 Prohibited equipment and appliance location. Equipment and appliances having an *ignition source* shall not be installed in Group H occupancies or control areas where open use, handling or dispensing of combustible, flammable or explosive materials occurs.

[FG] 304.5 Hydrogen-generating and refueling operations. Hydrogen-generating and refueling appliances shall be installed and located in accordance with their listing and the manufacturer's instructions. Ventilation shall be required in accordance with Section 304.5.1, 304.5.2 or 304.5.3 in public garages, private garages, repair garages, automotive motor fuel-dispensing facilities and parking garages that contain hydrogen-generating appliances or refueling systems. For the purpose of this section, rooms or spaces that are not part of the living space of a *dwelling unit* and that communicate directly with a private garage through openings shall be considered to be part of the private garage.

[FG] 304.5.1 Natural ventilation. Indoor locations intended for hydrogen-generating or refueling operations shall be limited to a maximum floor area of 850 square feet (79 m^2) and shall communicate with the outdoors in accordance with Sections 304.5.1.1 and 304.5.1.2. The maximum rated output capacity of hydrogen-generating appliances shall not exceed 4 standard cubic feet per minute (0.00189 m^3/s) of hydrogen for each 250 square feet (23.2 m^2) of floor area in such spaces. The minimum cross-sectional dimension of air openings shall be 3 inches (76 mm). Where ducts are used, they shall be of the same cross-sectional area as the free area of the openings to which they connect. In such locations, *equipment* and appliances having an *ignition source* shall be located such that the source of ignition is not within 12 inches (305 mm) of the ceiling.

[FG] 304.5.1.1 Two openings. Two permanent openings shall be provided within the garage. The upper opening shall be located entirely within 12 inches (305 mm) of the ceiling of the garage. The lower opening shall be located entirely within 12 inches (305 mm) of the floor of the garage. Both openings shall be provided in the same exterior wall. The openings shall communicate directly with the outdoors and shall have a minimum free area of $^1/_2$ square foot per 1,000 cubic feet (1 m^2/610 m^3) of garage volume.

[FG] 304.5.1.2 Louvers and grilles. In calculating free area required by Section 304.5.1, the required size of openings shall be based on the net free area of each opening. If the free area through a design of louver or grille is known, it shall be used in calculating the size opening required to provide the free area specified. If the design and free area are not known, it shall be assumed that wood louvers will have 25 percent free area and metal louvers and grilles will have 75 percent free area. Louvers and grilles shall be fixed in the open position.

[FG] 304.5.2 Mechanical ventilation. Indoor locations intended for hydrogen-generating or refueling operations shall be ventilated in accordance with Section 502.16. In such locations, *equipment* and appliances having an *ignition source* shall be located such that the source of ignition is below the mechanical ventilation outlet(s).

[FG] 304.5.3 Specially engineered installations. As an alternative to the provisions of Sections 304.5.1 and 304.5.2, the necessary supply of air for ventilation and dilution of flammable gases shall be provided by an *approved* engineered system.

304.6 Public garages. Appliances located in public garages, motor fueling-dispensing facilities, repair garages or other areas frequented by motor vehicles, shall be installed not less than 8 feet (2438 mm) above the floor. Where motor vehicles are capable of passing under an *appliance*, the *appliance* shall be installed at the clearances required by the *appliance* manufacturer and not less than 1 foot (305 mm) higher than the tallest vehicle garage door opening.

Exception: The requirements of this section shall not apply where the appliances are protected from motor vehicle impact and installed in accordance with Section 304.3 and NFPA 30A.

304.7 Private garages. Appliances located in private garages and carports shall be installed with a minimum clearance of 6 feet (1829 mm) above the floor.

Exception: The requirements of this section shall not apply where the appliances are protected from motor vehicle impact and installed in accordance with Section 304.3.

304.8 Construction and protection. Boiler rooms and furnace rooms shall be protected as required by the *International Building Code*.

304.9 Clearances to combustible construction. Heat-producing *equipment* and *appliances* shall be installed to maintain the required *clearances* to combustible construction as specified in the listing and manufacturer's instructions. Such clearances shall be reduced only in accordance with Section 308. *Clearances* to combustibles shall include such considerations as door swing, drawer pull, overhead projections or shelving and window swing, shutters, coverings and drapes. Devices such as doorstops or limits, closers, drapery ties or guards shall not be used to provide the required *clearances*.

304.10 Clearances from grade. Equipment and *appliances* installed at grade level shall be supported on a level concrete slab or other *approved* material extending not less than 3 inches (76 mm) above adjoining grade or shall be suspended not less than 6 inches (152 mm) above adjoining grade. Such support shall be in accordance with the manufacturer's installation instructions.

[BE] 304.11 Guards. Guards shall be provided where various components that require service and roof hatch openings are located within 10 feet (3048 mm) of a roof edge or open side of a walking surface and such edge or open side is located more than 30 inches (762 mm) above the floor, roof, or grade below. The guard shall extend not less than 30 inches (762 mm) beyond each end of components that require service. The top of the guard shall be located not less than 42 inches (1067 mm) above the elevated surface adjacent to the guard. The guard shall be constructed so as to prevent the passage of a 21-inch-diameter (533 mm) sphere and shall comply with the loading requirements for guards specified in the *International Building Code*.

Exception: Guards are not required where permanent fall arrest/restraint anchorage connector devices that comply with ANSI/ASSE Z 359.1 are affixed for use during the entire lifetime of the roof covering. The devices shall be re-evaluated for possible replacement when the entire roof covering is replaced. The devices shall be placed not more than 10 feet (3048 mm) on center along hip and ridge lines and placed not less than 10 feet (3048 mm) from roof edges and the open sides of walking surfaces.

304.12 Area served. Appliances serving different areas of a building other than where they are installed shall be permanently marked in an *approved* manner that uniquely identifies the *appliance* and the area it serves.

SECTION 305
PIPING SUPPORT

305.1 General. Mechanical system piping shall be supported in accordance with this section.

305.2 Materials. Pipe hangers and supports shall have sufficient strength to withstand all anticipated static and specified dynamic loading conditions associated with the intended use. Pipe hangers and supports that are in direct contact with piping shall be of *approved* materials that are compatible with the piping and that will not promote galvanic action.

305.3 Structural attachment. Hangers and anchors shall be attached to the building construction in an *approved* manner.

305.4 Interval of support. Piping shall be supported at distances not exceeding the spacing specified in Table 305.4, or in accordance with ANSI/MSS SP-58.

305.5 Protection against physical damage. In concealed locations where piping, other than cast-iron or steel, is installed through holes or notches in studs, joists, rafters or similar members less than $1^1/_2$ inches (38 mm) from the nearest edge of the member, the pipe shall be protected by shield plates. Protective steel shield plates having a minimum thickness of 0.0575 inch (1.463 mm) (No. 16 gage) shall cover the area of the pipe where the member is notched or bored, and shall extend not less than 2 inches (51 mm) above sole plates and below top plates.

SECTION 306
ACCESS AND SERVICE SPACE

306.1 Access. Appliances, controls devices, heat exchangers and HVAC system components that utilize energy shall be accessible for inspection, service, repair and replacement without disabling the function of a fire-resistance-rated assembly or removing permanent construction, other appliances, venting systems or any other piping or ducts not connected to the *appliance* being inspected, serviced, repaired or replaced. A level working space not less than 30 inches deep and 30 inches wide (762 mm by 762 mm) shall be provided in front of the control side to service an *appliance*.

306.1.1 Central furnaces. Central furnaces within compartments or alcoves shall have a minimum working space *clearance* of 3 inches (76 mm) along the sides, back and top with a total width of the enclosing space being not less than 12 inches (305 mm) wider than the furnace. Furnaces having a firebox open to the atmosphere shall have not less than 6 inches (152 mm) working space along the front

TABLE 305.4
PIPING SUPPORT SPACING[a]

PIPING MATERIAL	MAXIMUM HORIZONTAL SPACING (feet)	MAXIMUM VERTICAL SPACING (feet)
ABS pipe	4	10[c]
Aluminum pipe and tubing	10	15
Brass pipe	10	10
Brass tubing, $1^1/_4$-inch diameter and smaller	6	10
Brass tubing, $1^1/_2$-inch diameter and larger	10	10
Cast-iron pipe[b]	5	15
Copper or copper-alloy pipe	12	10
Copper or copper-alloy tubing, $1^1/_4$-inch diameter and smaller	6	10
Copper or copper-alloy tubing, $1^1/_2$-inch diameter and larger	10	10
CPVC pipe or tubing, 1 inch and smaller	3	10[c]
CPVC pipe or tubing, $1^1/_4$-inch and larger	4	10[c]
Lead pipe	Continuous	4
PB pipe or tubing	$2^2/_3$ (32 inches)	4
PE-RT 1 inch and smaller	$2^2/_3$ (32 inches)	10[c]
PE-RT $1^1/_4$ inches and larger	4	10[c]
PEX tubing	$2^2/_3$ (32 inches)	10[c]
Polypropylene (PP) pipe or tubing, 1 inch and smaller	$2^2/_3$ (32 inches)	10[c]
Polypropylene (PP) pipe or tubing, $1^1/_4$ inches and larger	4	10[c]
PVC pipe	4	10[c]
Steel tubing	8	10
Steel pipe	12	15

For SI: 1 inch = 25.4 mm, 1 foot = 304.8 mm.

a. See Section 301.18.

b. The maximum horizontal spacing of cast-iron pipe hangers shall be increased to 10 feet where 10-foot lengths of pipe are installed.

c. Mid-story guide.

combustion chamber side. *Combustion air* openings at the rear or side of the compartment shall comply with the requirements of Chapter 7.

Exception: This section shall not apply to replacement appliances installed in existing compartments and alcoves where the working space clearances are in accordance with the *equipment* or *appliance* manufacturer's installation instructions.

306.2 Appliances in rooms. Rooms containing appliances shall be provided with a door and an unobstructed passageway measuring not less than 36 inches (914 mm) wide and 80 inches (2032 mm) high.

Exception: Within a *dwelling unit*, appliances installed in a compartment, alcove, basement or similar space shall be accessed by an opening or door and an unobstructed passageway measuring not less than 24 inches (610 mm) wide and large enough to allow removal of the largest *appliance* in the space, provided that a level service space of not less than 30 inches (762 mm) deep and the height of the *appliance*, but not less than 30 inches (762 mm), is present at the front or service side of the *appliance* with the door open.

306.3 Appliances in attics. Attics containing appliances shall be provided with an opening and unobstructed passageway large enough to allow removal of the largest *appliance*. The passageway shall be not less than 30 inches (762 mm) high and 22 inches (559 mm) wide and not more than 20 feet (6096 mm) in length measured along the centerline of the passageway from the opening to the *appliance*. The passageway shall have continuous solid flooring not less than 24 inches (610 mm) wide. A level service space not less than 30 inches (762 mm) deep and 30 inches (762 mm) wide shall be present at the front or service side of the *appliance*. The clear access opening dimensions shall be not less than 20 inches by 30 inches (508 mm by 762 mm), and large enough to allow removal of the largest *appliance*.

Exceptions:

1. The passageway and level service space are not required where the *appliance* is capable of being serviced and removed through the required opening.

2. Where the passageway is unobstructed and not less than 6 feet (1829 mm) high and 22 inches (559 mm) wide for its entire length, the passageway shall be not greater than 50 feet (15 250 mm) in length.

306.3.1 Electrical requirements. A luminaire controlled by a switch located at the required passageway opening and a receptacle outlet shall be provided at or near the *appliance* location in accordance with NFPA 70.

306.4 Appliances under floors. Underfloor spaces containing appliances shall be provided with an access opening and unobstructed passageway large enough to remove the largest *appliance*. The passageway shall be not less than 30 inches (762 mm) high and 22 inches (559 mm) wide, nor more than 20 feet (6096 mm) in length measured along the centerline of the passageway from the opening to the *appliance*. A level service space not less than 30 inches (762 mm) deep and 30 inches (762 mm) wide shall be present at the front or service side of the *appliance*. If the depth of the passageway or the service space exceeds 12 inches (305 mm) below the adjoining grade, the walls of the passageway shall be lined with concrete or masonry. Such concrete or masonry shall extend not less than 4 inches (102 mm) above the adjoining grade and shall have sufficient lateral-bearing capacity to resist collapse. The clear access opening dimensions shall be not less

than 22 inches by 30 inches (559 mm by 762 mm), and large enough to allow removal of the largest *appliance*.

Exceptions:

1. The passageway is not required where the level service space is present when the access is open and the *appliance* is capable of being serviced and removed through the required opening.

2. Where the passageway is unobstructed and not less than 6 feet high (1929 mm) and 22 inches (559 mm) wide for its entire length, the passageway shall not be limited in length.

306.4.1 Electrical requirements. A luminaire controlled by a switch located at the required passageway opening and a receptacle outlet shall be provided at or near the *appliance* location in accordance with NFPA 70.

306.5 Equipment and appliances on roofs or elevated structures. Where *equipment* requiring access or appliances are located on an elevated structure or the roof of a building such that personnel will have to climb higher than 16 feet (4877 mm) above grade to access such equipment or appliances, an interior or exterior means of access shall be provided. Such access shall not require climbing over obstructions greater than 30 inches (762 mm) in height or walking on roofs having a slope greater than 4 units vertical in 12 units horizontal (33-percent slope). Such access shall not require the use of portable ladders. Where access involves climbing over parapet walls, the height shall be measured to the top of the parapet wall.

Permanent ladders installed to provide the required access shall comply with the following minimum design criteria:

1. The side railing shall extend above the parapet or roof edge not less than 30 inches (762 mm).

2. Ladders shall have rung spacing not to exceed 14 inches (356 mm) on center. The uppermost rung shall be not greater than 24 inches (610 mm) below the upper edge of the roof hatch, roof or parapet, as applicable.

3. Ladders shall have a toe spacing not less than 6 inches (152 mm) deep.

4. There shall be not less than 18 inches (457 mm) between rails.

5. Rungs shall have a diameter not less than 0.75-inch (19 mm) and be capable of withstanding a 300-pound (136.1 kg) load.

6. Ladders over 30 feet (9144 mm) in height shall be provided with offset sections and landings capable of withstanding 100 pounds per square foot (488.2 kg/m²). Landing dimensions shall be not less than 18 inches (457 mm) and not less than the width of the ladder served. A guard rail shall be provided on all open sides of the landing.

7. Climbing clearance. The distance from the centerline of the rungs to the nearest permanent object on the

climbing side of the ladder shall be not less than 30 inches (762 mm) measured perpendicular to the rungs. This distance shall be maintained from the point of ladder access to the bottom of the roof hatch. A minimum clear width of 15 inches (381 mm) shall be provided on both sides of the ladder measured from the midpoint of and parallel with the rungs except where cages or wells are installed.

8. Landing required. The ladder shall be provided with a clear and unobstructed bottom landing area having a minimum dimension of 30 inches (762 mm) by 30 inches (762 mm) centered in front of the ladder.

9. Ladders shall be protected against corrosion by *approved* means.

10. Access to ladders shall be provided at all times.

Catwalks installed to provide the required access shall be not less than 24 inches (610 mm) wide and shall have railings as required for service platforms.

Exception: This section shall not apply to Group R-3 occupancies.

306.5.1 Sloped roofs. Where appliances, *equipment*, fans or other components that require service are installed on a roof having a slope of three units vertical in 12 units horizontal (25-percent slope) or greater and having an edge more than 30 inches (762 mm) above grade at such edge, a level platform shall be provided on each side of the *appliance* or *equipment* to which access is required for service, repair or maintenance. The platform shall be not less than 30 inches (762 mm) in any dimension and shall be provided with guards. The guards shall extend not less than 42 inches (1067 mm) above the platform, shall be constructed so as to prevent the passage of a 21-inch-diameter (533 mm) sphere and shall comply with the loading requirements for guards specified in the *International Building Code*. Access shall not require walking on roofs having a slope greater than four units vertical in 12 units horizontal (33-percent slope). Where access involves obstructions greater than 30 inches (762 mm) in height, such obstructions shall be provided with ladders installed in accordance with Section 306.5 or stairways installed in accordance with the requirements specified in the *International Building Code* in the path of travel to and from appliances, fans or *equipment* requiring service.

306.5.2 Electrical requirements. A receptacle outlet shall be provided at or near the *equipment* location in accordance with NFPA 70.

SECTION 307
CONDENSATE DISPOSAL

307.1 Fuel-burning appliances. Liquid *combustion* by-products of condensing appliances shall be collected and discharged to an *approved* plumbing fixture or disposal area in accordance with the manufacturer's installation instructions. Condensate piping shall be of *approved* corrosion-resistant material and shall not be smaller than the drain connection on the appliance. Such piping shall maintain a minimum hori-

zontal slope in the direction of discharge of not less than one-eighth unit vertical in 12 units horizontal (1-percent slope).

307.2 Evaporators and cooling coils. Condensate drain systems shall be provided for *equipment* and appliances containing evaporators or cooling coils. Condensate drain systems shall be designed, constructed and installed in accordance with Sections 307.2.1 through 307.2.5.

Exception: Evaporators and cooling coils that are designed to operate in sensible cooling only and not support condensation shall not be required to meet the requirements of this section.

307.2.1 Condensate disposal. Condensate from all cooling coils and evaporators shall be conveyed from the drain pan outlet to an *approved* place of disposal. Such piping shall maintain a minimum horizontal slope in the direction of discharge of not less than one-eighth unit vertical in 12 units horizontal (1-percent slope). Condensate shall not discharge into a street, alley or other areas so as to cause a nuisance.

307.2.2 Drain pipe materials and sizes. Components of the condensate disposal system shall be cast iron, galvanized steel, copper, cross-linked polyethylene, polyethylene, ABS, CPVC, PVC, or polypropylene pipe or tubing. Components shall be selected for the pressure and temperature rating of the installation. Joints and connections shall be made in accordance with the applicable provisions of Chapter 7 of the *International Plumbing Code* relative to the material type. Condensate waste and drain line size shall be not less than $^3/_4$-inch (19.1 mm) internal diameter and shall not decrease in size from the drain pan connection to the place of condensate disposal. Where the drain pipes from more than one unit are manifolded together for condensate drainage, the pipe or tubing shall be sized in accordance with Table 307.2.2.

TABLE 307.2.2
CONDENSATE DRAIN SIZING

EQUIPMENT CAPACITY	MINIMUM CONDENSATE PIPE DIAMETER
Up to 20 tons of refrigeration	$^3/_4$ inch
Over 20 tons to 40 tons of refrigeration	1 inch
Over 40 tons to 90 tons of refrigeration	$1^1/_4$ inches
Over 90 tons to 125 tons of refrigeration	$1^1/_2$ inches
Over 125 tons to 250 tons of refrigeration	2 inches

1 inch = 25.4 mm, 1 ton = 3.517 kW.

307.2.3 Auxiliary and secondary drain systems. In addition to the requirements of Section 307.2.1, where damage to any building components could occur as a result of overflow from the *equipment* primary condensate removal system, one of the following auxiliary protection methods shall be provided for each cooling coil or fuel-fired *appliance* that produces condensate:

1. An auxiliary drain pan with a separate drain shall be provided under the coils on which condensation will occur. The auxiliary pan drain shall discharge to a conspicuous point of disposal to alert occupants in the event of a stoppage of the primary drain. The pan

shall have a minimum depth of $1^1/_2$ inches (38 mm), shall be not less than 3 inches (76 mm) larger than the unit, or the coil dimensions in width and length and shall be constructed of corrosion-resistant material. Galvanized sheet steel pans shall have a minimum thickness of not less than 0.0236 inch (0.6010 mm) (No. 24 gage). Nonmetallic pans shall have a minimum thickness of not less than 0.0625 inch (1.6 mm).

2. A separate overflow drain line shall be connected to the drain pan provided with the *equipment*. Such overflow drain shall discharge to a conspicuous point of disposal to alert occupants in the event of a stoppage of the primary drain. The overflow drain line shall connect to the drain pan at a higher level than the primary drain connection.

3. An auxiliary drain pan without a separate drain line shall be provided under the coils on which condensate will occur. Such pan shall be equipped with a water-level detection device conforming to UL 508 that will shut off the *equipment* served prior to overflow of the pan. The auxiliary drain pan shall be constructed in accordance with Item 1 of this section.

4. A water-level detection device conforming to UL 508 shall be provided that will shut off the *equipment* served in the event that the primary drain is blocked. The device shall be installed in the primary drain line, the overflow drain line, or in the equipment-supplied drain pan, located at a point higher than the primary drain line connection and below the overflow rim of such pan.

Exception: Fuel-fired appliances that automatically shut down operation in the event of a stoppage in the condensate drainage system.

307.2.3.1 Water-level monitoring devices. On downflow units and all other coils that do not have a secondary drain or provisions to install a secondary or auxiliary drain pan, a water-level monitoring device shall be installed inside the primary drain pan. This device shall shut off the *equipment* served in the event that the primary drain becomes restricted. Devices installed in the drain line shall not be permitted.

307.2.3.2 Appliance, equipment and insulation in pans. Where appliances, *equipment* or insulation are subject to water damage when auxiliary drain pans fill, that portion of the *appliance*, *equipment* and insulation shall be installed above the rim of the pan. Supports located inside of the pan to support the *appliance* or *equipment* shall be water resistant and *approved*.

307.2.4 Traps. Condensate drains shall be trapped as required by the *equipment* or *appliance* manufacturer.

307.2.4.1 Ductless mini-split system traps. Ductless mini-split equipment that produces condensate shall be provided with an inline check valve located in the drain line, or a trap.

307.2.5 Drain line maintenance. Condensate drain lines shall be configured to permit the clearing of blockages and performance of maintenance without requiring the drain line to be cut.

307.3 Condensate pumps. Condensate pumps located in uninhabitable spaces, such as attics and crawl spaces, shall be connected to the appliance or equipment served such that when the pump fails, the appliance or equipment will be prevented from operating. Pumps shall be installed in accordance with the manufacturers' instructions.

SECTION 308
CLEARANCE REDUCTION

308.1 Scope. This section shall govern the reduction in required *clearances* to combustible materials and combustible assemblies for *chimneys*, vents, kitchen exhaust equipment, mechanical appliances, and mechanical devices and *equipment*.

308.2 Listed appliances and equipment. The reduction of the required *clearances* to combustibles for *listed* and *labeled* appliances and *equipment* shall be in accordance with the requirements of this section except that such clearances shall not be reduced where reduction is specifically prohibited by the terms of the *appliance* or *equipment* listing.

308.3 Protective assembly construction and installation. Reduced *clearance* protective assemblies, including structural and support elements, shall be constructed of noncombustible materials. Spacers utilized to maintain an airspace between the protective assembly and the protected material or assembly shall be noncombustible. Where a space between the protective assembly and protected combustible material or assembly is specified, the same space shall be provided around the edges of the protective assembly and the spacers shall be placed so as to allow air circulation by convection in such space. Protective assemblies shall not be placed less than 1 inch (25 mm) from the mechanical appliances, devices or *equipment*, regardless of the allowable reduced *clearance*.

308.4 Allowable reduction. The reduction of required *clearances* to combustible assemblies or combustible materials shall be based on the utilization of a reduced *clearance* protective assembly in accordance with Section 308.4.1 or 308.4.2.

308.4.1 Labeled assemblies. The allowable clearance reduction shall be based on an approved reduced clearance protective assembly that is listed and labeled in accordance with UL 1618.

308.4.2 Reduction table. The allowable *clearance* reduction shall be based on one of the methods specified in Table 308.4.2. Where required *clearances* are not listed in Table 308.4.2, the reduced *clearances* shall be determined by linear interpolation between the distances listed in the table. Reduced *clearances* shall not be derived by extrapolation below the range of the table.

308.4.2.1 Solid fuel-burning appliances. The *clearance* reduction methods specified in Table 308.4.2 shall not be utilized to reduce the *clearance* required for solid fuel-burning appliances that are *labeled* for installation with clearances of 12 inches (305 mm) or less. Where appliances are *labeled* for installation with *clearances* of greater than 12 inches (305 mm), the *clearance* reduction methods of Table 308.4.2 shall not reduce the *clearance* to less than 12 inches (305 mm).

308.4.2.2 Masonry chimneys. The *clearance* reduction methods specified in Table 308.4.2 shall not be utilized to reduce the *clearances* required for masonry *chimneys* as specified in Chapter 8 and the *International Building Code*.

308.4.2.3 Chimney connector pass-throughs. The *clearance* reduction methods specified in Table 308.4.2 shall not be utilized to reduce the clearances required for *chimney* connector pass-throughs as specified in Section 803.10.4.

308.4.2.4 Masonry fireplaces. The *clearance* reduction methods specified in Table 308.4.2 shall not be utilized to reduce the clearances required for masonry fireplaces as specified in Chapter 8 and the *International Building Code*.

308.4.2.5 Kitchen exhaust ducts. The *clearance* reduction methods specified in Table 308.4.2 shall not be utilized to reduce the minimum *clearances* required by Section 506.3.11.1 for kitchen exhaust ducts enclosed in a shaft.

SECTION 309
TEMPERATURE CONTROL

[BG] 309.1 Space-heating systems. Interior spaces intended for human occupancy shall be provided with active or passive space-heating systems capable of maintaining an indoor temperature of not less than 68°F (20°C) at a point 3 feet (914 mm) above floor on the design heating day. The installation of portable space heaters shall not be used to achieve compliance with this section.

Exceptions:

1. Interior spaces where the primary purpose is not associated with human comfort.

2. Group F, H, S and U occupancies.

SECTION 310
EXPLOSION CONTROL

[F] 310.1 Required. Structures occupied for purposes involving explosion hazards shall be provided with explosion control where required by the *International Fire Code*. Explosion control systems shall be designed and installed in accordance with Section 911 of the *International Fire Code*.

TABLE 308.4.2
CLEARANCE REDUCTION METHODS[b]

TYPE OF PROTECTIVE ASSEMBLY[a]	REDUCED CLEARANCE WITH PROTECTION (inches)[a]							
	Horizontal combustible assemblies located above the heat source				Horizontal combustible assemblies located beneath the heat source and all vertical combustible assemblies			
	Required clearance to combustibles without protection (inches)[a]				Required clearance to combustibles without protection (inches)			
	36	18	9	6	36	18	9	6
Galvanized sheet steel, having a minimum thickness of 0.0236 inch (No. 24 gage), mounted on 1-inch glass fiber or mineral wool batt reinforced with wire on the back, 1 inch off the combustible assembly	18	9	5	3	12	6	3	3
Galvanized sheet steel, having a minimum thickness of 0.0236 inch (No. 24 gage), spaced 1 inch off the combustible assembly	18	9	5	3	12	6	3	2
Two layers of galvanized sheet steel, having a minimum thickness of 0.0236 inch (No. 24 gage), having a 1-inch airspace between layers, spaced 1 inch off the combustible assembly	18	9	5	3	12	6	3	3
Two layers of galvanized sheet steel, having a minimum thickness of 0.0236 inch (No. 24 gage), having 1 inch of fiberglass insulation between layers, spaced 1 inch off the combustible assembly	18	9	5	3	12	6	3	3
0.5-inch inorganic insulating board, over 1 inch of fiberglass or mineral wool batt, against the combustible assembly	24	12	6	4	18	9	5	3
3¹/₂-inch brick wall, spaced 1 inch off the combustible wall	—	—	—	—	12	6	6	6
3¹/₂-inch brick wall, against the combustible wall	—	—	—	—	24	12	6	5

For SI: 1 inch = 25.4 mm, °C = [(°F)-32]/1.8, 1 pound per cubic foot = 16.02 kg/m³, 1.0 Btu • in/(ft² • h • °F) = 0.144 W/m² • K.

a. Mineral wool and glass fiber batts (blanket or board) shall have a minimum density of 8 pounds per cubic foot and a minimum melting point of 1,500°F. Insulation material utilized as part of a clearance reduction system shall have a thermal conductivity of 1.0 Btu • in/(ft² • h • °F) or less. Insulation board shall be formed of noncombustible material.

b. For limitations on clearance reduction for solid fuel-burning appliances, masonry chimneys, connector pass-throughs, masonry fire places and kitchen ducts, see Sections 308.4.2.1 through 308.4.2.5.

SECTION 311
SMOKE AND HEAT VENTS

[F] 311.1 Required. *Approved* smoke and heat vents shall be installed in the roofs of one-story buildings where required by the *International Fire Code*. Smoke and heat vents shall be designed and installed in accordance with the *International Fire Code*.

SECTION 312
HEATING AND COOLING LOAD CALCULATIONS

312.1 Load calculations. Heating and cooling system design loads for the purpose of sizing systems, appliances and *equipment* shall be determined in accordance with the procedures described in the ASHRAE/ACCA Standard 183. Alternatively, design loads shall be determined by an *approved* equivalent computation procedure, using the design parameters specified in Chapter 3 [CE] of the *International Energy Conservation Code*.

CHAPTER 4

VENTILATION

SECTION 401
GENERAL

401.1 Scope. This chapter shall govern the ventilation of spaces within a building intended to be occupied. Mechanical exhaust systems, including exhaust systems serving clothes dryers and cooking appliances; hazardous exhaust systems; dust, stock and refuse conveyor systems; subslab soil exhaust systems; smoke control systems; energy recovery ventilation systems and other systems specified in Section 502 shall comply with Chapter 5.

401.2 Ventilation required. Every occupied space shall be ventilated by natural means in accordance with Section 402 or by mechanical means in accordance with Section 403. Where the air infiltration rate in a dwelling unit is less than 5 air changes per hour when tested with a blower door at a pressure of 0.2-inch water column (50 Pa) in accordance with Section R402.4.1.2 of the *International Energy Conservation Code*, the dwelling unit shall be ventilated by mechanical means in accordance with Section 403. Ambulatory care facilities and Group I-2 occupancies shall be ventilated by mechanical means in accordance with Section 407.

401.3 When required. Ventilation shall be provided during the periods that the room or space is occupied.

401.4 Intake opening location. Air intake openings shall comply with all of the following:

1. Intake openings shall be located not less than 10 feet (3048 mm) from lot lines or buildings on the same lot.

2. Mechanical and gravity outdoor air intake openings shall be located not less than 10 feet (3048 mm) horizontally from any hazardous or noxious contaminant source, such as vents, streets, alleys, parking lots and loading docks, except as specified in Item 3 or Section 501.3.1. Outdoor air intake openings shall be permitted to be located less than 10 feet (3048 mm) horizontally from streets, alleys, parking lots and loading docks provided that the openings are located not less than 25 feet (7620 mm) vertically above such locations. Where openings front on a street or public way, the distance shall be measured from the closest edge of the street or public way.

3. Intake openings shall be located not less than 3 feet (914 mm) below contaminant sources where such sources are located within 10 feet (3048 mm) of the opening.

4. Intake openings on structures in flood hazard areas shall be at or above the elevation required by Section 1612 of the *International Building Code* for utilities and attendant equipment.

401.5 Intake opening protection. Air intake openings that terminate outdoors shall be protected with corrosion-resistant screens, louvers or grilles. Openings in louvers, grilles and screens shall be sized in accordance with Table 401.5, and shall be protected against local weather conditions. Louvers that protect air intake openings in structures located in hurricane-prone regions, as defined in the *International Building Code*, shall comply with AMCA 550. Outdoor air intake openings located in exterior walls shall meet the provisions for exterior wall opening protectives in accordance with the *International Building Code*.

TABLE 401.5
OPENING SIZES IN LOUVERS, GRILLES AND SCREENS PROTECTING AIR INTAKE OPENINGS

OUTDOOR OPENING TYPE	MINIMUM AND MAXIMUM OPENING SIZES IN LOUVERS, GRILLES AND SCREENS MEASURED IN ANY DIRECTION
Intake openings in residential occupancies	Not $< \frac{1}{4}$ inch and not $> \frac{1}{2}$ inch
Intake openings in other than residential occupancies	$> \frac{1}{4}$ inch and not > 1 inch

For SI: 1 inch = 25.4 mm.

401.6 Contaminant sources. Stationary local sources producing air-borne particulates, heat, odors, fumes, spray, vapors, smoke or gases in such quantities as to be irritating or injurious to health shall be provided with an exhaust system in accordance with Chapter 5 or a means of collection and removal of the contaminants. Such exhaust shall discharge directly to an *approved* location at the exterior of the building.

SECTION 402
NATURAL VENTILATION

[BG] 402.1 Natural ventilation. *Natural ventilation* of an occupied space shall be through windows, doors, louvers or other openings to the outdoors. The operating mechanism for such openings shall be provided with ready access so that the openings are readily controllable by the building occupants.

[BG] 402.2 Ventilation area required. The minimum openable area to the outdoors shall be 4 percent of the floor area being ventilated.

[BG] 402.3 Adjoining spaces. Where rooms and spaces without openings to the outdoors are ventilated through an adjoining room, the opening to the adjoining rooms shall be unobstructed and shall have an area not less than 8 percent of the floor area of the interior room or space, but not less than 25 square feet (2.3 m²). The minimum openable area to the outdoors shall be based on the total floor area being ventilated.

Exception: Exterior openings required for ventilation shall be permitted to open into a thermally isolated sunroom addition or patio cover, provided that the openable area between the sunroom addition or patio cover and the interior room has an area of not less than 8 percent of the floor area of the interior room or space, but not less than 20 square feet (1.86 m²). The minimum openable area to the outdoors shall be based on the total floor area being ventilated.

[BG] 402.4 Openings below grade. Where openings below grade provide required *natural ventilation*, the outside horizontal clear space measured perpendicular to the opening shall be one and one-half times the depth of the opening. The depth of the opening shall be measured from the average adjoining ground level to the bottom of the opening.

SECTION 403
MECHANICAL VENTILATION

403.1 Ventilation system. Mechanical ventilation shall be provided by a method of supply air and return or *exhaust air* except that mechanical ventilation air requirements for Group R-2, R-3 and R-4 occupancies three stories and less in height above grade plane shall be provided by an exhaust system, supply system or combination thereof. The amount of supply air shall be approximately equal to the amount of return and *exhaust air*. The system shall not be prohibited from producing negative or positive pressure. The system to convey *ventilation air* shall be designed and installed in accordance with Chapter 6.

403.2 Outdoor air required. The minimum outdoor airflow rate shall be determined in accordance with Section 403.3.

> **Exception:** Where the *registered design professional* demonstrates that an engineered ventilation system design will prevent the maximum concentration of contaminants from exceeding that obtainable by the rate of outdoor air ventilation determined in accordance with Section 403.3, the minimum required rate of outdoor air shall be reduced in accordance with such engineered system design.

403.2.1 Recirculation of air. The outdoor air required by Section 403.3 shall not be recirculated. Air in excess of that required by Section 403.3 shall not be prohibited from being recirculated as a component of supply air to building spaces, except that:

1. Ventilation air shall not be recirculated from one *dwelling* to another or to dissimilar occupancies.

2. Supply air to a swimming pool and associated deck areas shall not be recirculated unless such air is dehumidified to maintain the relative humidity of the area at 60 percent or less. Air from this area shall not be recirculated to other spaces where more than 10 percent of the resulting supply airstream consists of air recirculated from these spaces.

3. Where mechanical exhaust is required by Note b in Table 403.3.1.1, recirculation of air from such spaces shall be prohibited. Recirculation of air that is contained completely within such spaces shall not be prohibited. Where recirculation of air is prohibited, all air supplied to such spaces shall be

exhausted, including any air in excess of that required by Table 403.3.1.1.

4. Where mechanical exhaust is required by Note g in Table 403.3.1.1, mechanical exhaust is required and recirculation from such spaces is prohibited where more than 10 percent of the resulting supply airstream consists of air recirculated from these spaces. Recirculation of air that is contained completely within such spaces shall not be prohibited.

403.2.2 Transfer air. Except where recirculation from such spaces is prohibited by Table 403.3.1.1, air transferred from occupiable spaces is not prohibited from serving as *makeup air* for required exhaust systems in such spaces as kitchens, baths, toilet rooms, elevators and smoking lounges. The amount of transfer air and *exhaust air* shall be sufficient to provide the flow rates as specified in Section 403.3.1.1. The required outdoor airflow rates specified in Table 403.3.1.1 shall be introduced directly into such spaces or into the occupied spaces from which air is transferred or a combination of both.

403.3 Outdoor air and local exhaust airflow rates. Group R-2, R-3 and R-4 occupancies three stories and less in height above grade plane shall be provided with outdoor air and local exhaust in accordance with Section 403.3.2. All other buildings intended to be occupied shall be provided with outdoor air and local exhaust in accordance with Section 403.3.1.

403.3.1 Other buildings intended to be occupied. The design of local exhaust systems and ventilation systems for outdoor air for occupancies other than Group R-2, R-3 and R-4 three stories and less above grade plane shall comply with Sections 403.3.1.1 through 403.3.1.5.

403.3.1.1 Outdoor airflow rate. Ventilation systems shall be designed to have the capacity to supply the minimum outdoor airflow rate, determined in accordance with this section. In each occupiable space, the ventilation system shall be designed to deliver the required rate of outdoor airflow to the *breathing zone*. The occupant load utilized for design of the ventilation system shall be not less than the number determined from the estimated maximum occupant load rate indicated in Table 403.3.1.1. Ventilation rates for occupancies not represented in Table 403.3.1.1 shall be those for a listed *occupancy* classification that is most similar in terms of occupant density, activities and building construction; or shall be determined by an *approved* engineering analysis. The ventilation system shall be designed to supply the required rate of *ventilation air* continuously during the period the building is occupied, except as otherwise stated in other provisions of the code.

With the exception of smoking lounges, the ventilation rates in Table 403.3.1.1 are based on the absence of smoking in occupiable spaces. Where smoking is anticipated in a space other than a smoking lounge, the ventilation system serving the space shall be designed to provide ventilation over and above that required by Table 403.3.1.1 in accordance with accepted engineering practice.

Exception: The occupant load is not required to be determined based on the estimated maximum occupant load rate indicated in Table 403.3.1.1 where *approved* statistical data document the accuracy of an alternate anticipated occupant density.

403.3.1.1.1 Zone outdoor airflow. The minimum outdoor airflow required to be supplied to each zone shall be determined as a function of *occupancy* classification and space air distribution effectiveness in accordance with Sections 403.3.1.1.1.1 through 403.3.1.1.1.3.

403.3.1.1.1.1 Breathing zone outdoor airflow. The outdoor airflow rate required in the *breathing zone* (V_{bz}) of the *occupiable space* or spaces in a zone shall be determined in accordance with Equation 4-1.

$$V_{bz} = R_p P_z + R_a A_z \qquad \textbf{(Equation 4-1)}$$

where:

A_z = Zone floor area: the *net occupiable floor area* of the space or spaces in the zone.

P_z = Zone population: the number of people in the space or spaces in the zone.

R_p = People outdoor air rate: the outdoor airflow rate required per person from Table 403.3.1.1.

R_a = Area outdoor air rate: the outdoor airflow rate required per unit area from Table 403.3.1.1.

403.3.1.1.1.2 Zone air distribution effectiveness. The zone air distribution effectiveness (E_z) shall be determined using Table 403.3.1.1.1.2.

TABLE 403.3.1.1.1.2
ZONE AIR DISTRIBUTION EFFECTIVENESS[a,b,c,d]

AIR DISTRIBUTION CONFIGURATION	E_z
Ceiling or floor supply of cool air	1.0[e]
Ceiling or floor supply of warm air and floor return	1.0
Ceiling supply of warm air and ceiling return	0.8[f]
Floor supply of warm air and ceiling return	0.7
Makeup air drawn in on the opposite side of the room from the exhaust and/or return	0.8
Makeup air drawn in near to the exhaust and/or return location	0.5

For SI: 1 foot = 304.8 mm, 1 foot per minute = 0.00506 m/s,
 °C = [(°F) – 32]/1.8.

a. "Cool air" is air cooler than space temperature.
b. "Warm air" is air warmer than space temperature.
c. "Ceiling" includes any point above the breathing zone.
d. "Floor" includes any point below the breathing zone.

e. Zone air distribution effectiveness of 1.2 shall be permitted for systems with a floor supply of cool air and ceiling return, provided that low-velocity displacement ventilation achieves unidirectional flow and thermal stratification.

f. Zone air distribution effectiveness of 1.0 shall be permitted for systems with a ceiling supply of warm air, provided that supply air temperature is less than 15°F above space temperature and provided that the 150-foot-per-minute supply air jet reaches to within $4^1/_2$ feet of floor level.

403.3.1.1.1.3 Zone outdoor airflow. The zone outdoor airflow rate (V_{oz}), shall be determined in accordance with Equation 4-2.

$$V_{oz} = \frac{V_{bz}}{E_z} \qquad \textbf{(Equation 4-2)}$$

403.3.1.1.2 System outdoor airflow. The outdoor air required to be supplied by each ventilation system shall be determined in accordance with Sections 403.3.1.1.2.1 through 403.3.1.1.2.3 as a function of system type and zone outdoor airflow rates.

403.3.1.1.2.1 Single zone systems. Where one air handler supplies a mixture of outdoor air and recirculated return air to only one zone, the system outdoor air intake flow rate (V_{ot}) shall be determined in accordance with Equation 4-3.

$$V_{ot} = V_{oz} \qquad \textbf{(Equation 4-3)}$$

403.3.1.1.2.2 100-percent outdoor air systems. Where one air handler supplies only outdoor air to one or more zones, the system outdoor air intake flow rate (V_{ot}) shall be determined using Equation 4-4.

$$V_{ot} = \Sigma_{all\,zones} V_{oz} \qquad \textbf{(Equation 4-4)}$$

403.3.1.1.2.3 Multiple zone recirculating systems. Where one air handler supplies a mixture of outdoor air and recirculated return air to more than one zone, the system outdoor air intake flow rate (V_{ot}) shall be determined in accordance with Sections 403.3.1.1.2.3.1 through 403.3.1.1.2.3.4.

403.3.1.1.2.3.1 Primary outdoor air fraction. The primary outdoor air fraction (Z_p) shall be determined for each zone in accordance with Equation 4-5.

$$Z_p = \frac{V_{oz}}{V_{pz}} \qquad \textbf{(Equation 4-5)}$$

where:

V_{pz} = Primary airflow: The airflow rate supplied to the zone from the air-handling unit at which the outdoor air intake is located. It includes outdoor intake air and recirculated air from that air-handling unit but does not include air transferred or air recirculated to the zone by other means. For design purposes, V_{pz} shall be the zone design primary airflow rate, except for zones with variable air volume supply and V_{pz} shall be the lowest expected primary airflow rate to the zone when it is fully occupied.

403.3.1.1.2.3.2 System ventilation efficiency.

The system ventilation efficiency (E_v) shall be determined using Table 403.3.1.1.2.3.2 or Appendix A of ASHRAE 62.1.

TABLE 403.3.1.1.2.3.2
SYSTEM VENTILATION EFFICIENCY[a,b]

Max (Z_p)	E_v
≤ 0.15	1
≤ 0.25	0.9
≤ 0.35	0.8
≤ 0.45	0.7
≤ 0.55	0.6
≤ 0.65	0.5
≤ 0.75	0.4
> 0.75	0.3

a. *Max* (Z_p) is the largest value of Z_p calculated using Equation 4-5 among all the zones served by the system.

b. Interpolating between table values shall be permitted.

403.3.1.1.2.3.3 Uncorrected outdoor air intake.

The uncorrected outdoor air intake flow rate (V_{ou}) shall be determined in accordance with Equation 4-6.

$$V_{ou} = D \Sigma_{all\,zones} R_p P_z + \Sigma_{all\,zones} R_a A_z$$

(Equation 4-6)

where:

D = Occupant diversity: the ratio of the system population to the sum of the zone populations, determined in accordance with Equation 4-7.

$$D = \frac{P_s}{\Sigma_{all\,zones} P_z} \qquad \textbf{(Equation 4-7)}$$

where:

P_s = System population: The total number of occupants in the area served by the system. For design purposes, P_s shall be the maximum number of occupants expected to be concurrently in all zones served by the system.

403.3.1.1.2.3.4 Outdoor air intake flow rate.

The outdoor air intake flow rate (V_{ot}) shall be determined in accordance with Equation 4-8.

$$V_{ot} = \frac{V_{ou}}{E_v} \qquad \textbf{(Equation 4-8)}$$

403.3.1.2 Exhaust ventilation.

Exhaust airflow rate shall be provided in accordance with the requirements of Table 403.3.1.1. Outdoor air introduced into a space by an exhaust system shall be considered as contributing to the outdoor airflow required by Table 403.3.1.1.

403.3.1.3 System operation.

The minimum flow rate of outdoor air that the ventilation system must be capable of supplying during its operation shall be permitted to be based on the rate per person indicated in Table 403.3.1.1 and the actual number of occupants present.

403.3.1.4 Variable air volume system control.

Variable air volume air distribution systems, other than those designed to supply only 100-percent outdoor air, shall be provided with controls to regulate the flow of outdoor air. Such control system shall be designed to maintain the flow rate of outdoor air at a rate of not less than that required by Section 403.3 over the entire range of supply air operating rates.

403.3.1.5 Balancing.

The *ventilation* air distribution system shall be provided with means to adjust the system to achieve not less than the minimum ventilation airflow rate as required by Sections 403.3 and 403.3.1.2. Ventilation systems shall be balanced by an *approved* method. Such balancing shall verify that the ventilation system is capable of supplying and exhausting the airflow rates required by Sections 403.3 and 403.3.1.2.

403.3.2 Group R-2, R-3 and R-4 occupancies, three stories and less.

The design of local exhaust systems and ventilation systems for outdoor air in Group R-2, R-3 and R-4 occupancies three stories and less in height above grade plane shall comply with Sections 403.3.2.1 through 403.3.2.3.

403.3.2.1 Outdoor air for dwelling units.

An outdoor air ventilation system consisting of a mechanical exhaust system, supply system or combination thereof shall be installed for each dwelling unit. Local exhaust or supply systems, including outdoor air ducts connected to the return side of an air handler, are permitted to serve as such a system. The outdoor air ventilation system shall be designed to provide the required rate of outdoor air continuously during the period that the building is occupied. The minimum continuous outdoor airflow rate shall be determined in accordance with Equation 4-9.

$$Q_{OA} = 0.01 A_{floor} + 7.5(N_{br} + 1) \qquad \textbf{(Equation 4-9)}$$

where:

Q_{OA} = outdoor airflow rate, cfm

A_{floor} = floor area, ft^2

N_{br} = number of bedrooms; not to be less than one

Exception: The outdoor air ventilation system is not required to operate continuously where the system has controls that enable operation for not less than 1 hour of each 4-hour period. The average outdoor air flow rate over the 4-hour period shall be not less than that prescribed by Equation 4-9.

403.3.2.2 Outdoor air for other spaces. Corridors and other common areas within the conditioned space shall be provided with outdoor air at a rate of not less than 0.06 cfm per square foot of floor area.

403.3.2.3 Local exhaust. Local exhaust systems shall be provided in kitchens, bathrooms and toilet rooms and shall have the capacity to exhaust the minimum airflow rate determined in accordance with Table 403.3.2.3.

TABLE 403.3.2.3
MINIMUM REQUIRED LOCAL EXHAUST RATES
FOR GROUP R-2, R-3, AND R-4 OCCUPANCIES

AREA TO BE EXHAUSTED	EXHAUST RATE CAPACITY
Kitchens	100 cfm intermittent or 25 cfm continuous
Bathrooms and toilet rooms	50 cfm intermittent or 20 cfm continuous

For SI: 1 cubic foot per minute = 0.0004719 m³/s.

SECTION 404
ENCLOSED PARKING GARAGES

404.1 Enclosed parking garages. Where mechanical ventilation systems for enclosed parking garages operate intermittently, such operation shall be automatic by means of carbon monoxide detectors applied in conjunction with nitrogen dioxide detectors. Such detectors shall be installed in accordance with their manufacturers' recommendations.

404.2 Minimum ventilation. Automatic operation of the system shall not reduce the ventilation airflow rate below 0.05 cfm per square foot ($0.00025 \text{ m}^3/\text{s} \cdot \text{m}^2$) of the floor area and the system shall be capable of producing a ventilation airflow rate of 0.75 cfm per square foot ($0.0038 \text{ m}^3/\text{s} \cdot \text{m}^2$) of floor area.

404.3 Occupied spaces accessory to public garages. Connecting offices, waiting rooms, ticket booths and similar uses that are accessory to a public garage shall be maintained at a positive pressure and shall be provided with ventilation in accordance with Section 403.3.1.

SECTION 405
SYSTEMS CONTROL

405.1 General. Mechanical ventilation systems shall be provided with manual or automatic controls that will operate such systems whenever the spaces are occupied. Air-conditioning systems that supply required *ventilation air* shall be provided with controls designed to automatically maintain the required outdoor air supply rate during occupancy.

SECTION 406
VENTILATION OF UNINHABITED SPACES

406.1 General. Uninhabited spaces, such as crawl spaces and attics, shall be provided with *natural ventilation* openings as required by the *International Building Code* or shall be pro-

vided with a mechanical exhaust and supply air system. The mechanical exhaust rate shall be not less than 0.02 cfm per square foot ($0.00001 \text{ m}^3/\text{s} \cdot \text{m}^2$) of horizontal area and shall be automatically controlled to operate when the relative humidity in the space served exceeds 60 percent.

SECTION 407
AMBULATORY CARE FACILITIES AND
GROUP I-2 OCCUPANCIES

407.1 General. Mechanical ventilation for ambulatory care facilities and Group I-2 occupancies shall be designed and installed in accordance with this code and ASHRAE 170.

TABLE 403.3.1.1
MINIMUM VENTILATION RATES

OCCUPANCY CLASSIFICATION	OCCUPANT DENSITY #/1000 FT² ᵃ	PEOPLE OUTDOOR AIRFLOW RATE IN BREATHING ZONE, R_p CFM/PERSON	AREA OUTDOOR AIRFLOW RATE IN BREATHING ZONE, R_a CFM/FT² ᵃ	EXHAUST AIRFLOW RATE CFM/FT² ᵃ
Correctional facilities				
Booking/waiting	50	7.5	0.06	—
Cells				
without plumbing fixtures	25	5	0.12	—
with plumbing fixtures[g]	25	5	0.12	1.0
Day room	30	5	0.06	—
Dining halls (see food and beverage service)	—	—	—	—
Guard stations	15	5	0.06	—
Dry cleaners, laundries				
Coin-operated dry cleaner	20	15	—	—
Coin-operated laundries	20	7.5	0.06	—
Commercial dry cleaner	30	30	—	—
Commercial laundry	10	25	—	—
Storage, pick up	30	7.5	0.12	—
Education				
Art classroom[g]	20	10	0.18	0.7
Auditoriums	150	5	0.06	—
Classrooms (ages 5-8)	25	10	0.12	—
Classrooms (age 9 plus)	35	10	0.12	—
Computer lab	25	10	0.12	—
Corridors (see public spaces)	—	—	—	—
Day care (through age 4)	25	10	0.18	—
Lecture classroom	65	7.5	0.06	—
Lecture hall (fixed seats)	150	7.5	0.06	—
Locker/dressing rooms[g]	—	—	—	0.25
Media center	25	10	0.12	—
Multiuse assembly	100	7.5	0.06	—
Music/theater/dance	35	10	0.06	—
Science laboratories[g]	25	10	0.18	1.0
Smoking lounges[b]	70	60	—	—
Sports locker rooms[g]	—	—	—	0.5
Wood/metal shops[g]	20	10	0.18	0.5
Food and beverage service				
Bars, cocktail lounges	100	7.5	0.18	—
Cafeteria, fast food	100	7.5	0.18	—
Dining rooms	70	7.5	0.18	—
Kitchens (cooking)[b]	—	—	—	0.7

(continued)

2015 INTERNATIONAL MECHANICAL CODE®

TABLE 403.3.1.1—continued
MINIMUM VENTILATION RATES

OCCUPANCY CLASSIFICATION	OCCUPANT DENSITY #/1000 FT² ª	PEOPLE OUTDOOR AIRFLOW RATE IN BREATHING ZONE, R_p CFM/PERSON	AREA OUTDOOR AIRFLOW RATE IN BREATHING ZONE, R_a CFM/FT² ª	EXHAUST AIRFLOW RATE CFM/FT² ª
Hotels, motels, resorts and dormitories				
Bathrooms/toilet—private[g]	—	—	—	25/50[f]
Bedroom/living room	10	5	0.06	—
Conference/meeting	50	5	0.06	—
Dormitory sleeping areas	20	5	0.06	—
Gambling casinos	120	7.5	0.18	—
Lobbies/prefunction	30	7.5	0.06	—
Multipurpose assembly	120	5	0.06	—
Offices				
Conference rooms	50	5	0.06	—
Main entry lobbies	10	5	0.06	—
Office spaces	5	5	0.06	—
Reception areas	30	5	0.06	—
Telephone/data entry	60	5	0.06	—
Private dwellings, single and multiple				
Garages, common for multiple units[b]	—	—	—	0.75
Kitchens[b]	—	—	—	25/100[f]
Living areas[c]	Based upon number of bedrooms. First bedroom, 2; each additional bedroom, 1	0.35 ACH but not less than 15 cfm/person	—	—
Toilet rooms and bathrooms[g]	—	—	—	20/50[f]
Public spaces				
Corridors	—	—	0.06	—
Courtrooms	70	5	0.06	—
Elevator car	—	—	—	1.0
Legislative chambers	50	5	0.06	—
Libraries	10	5	0.12	—
Museums (children's)	40	7.5	0.12	—
Museums/galleries	40	7.5	0.06	—
Places of religious worship	120	5	0.06	—
Shower room (per shower head)[g]	—	—	—	50/20[f]
Smoking lounges[b]	70	60	—	—
Toilet rooms — public[g]	—	—	—	50/70[e]
Retail stores, sales floors and showroom floors				
Dressing rooms	—	—	—	0.25
Mall common areas	40	7.5	0.06	—
Sales	15	7.5	0.12	—
Shipping and receiving	—	—	0.12	—
Smoking lounges[b]	70	60	—	—
Storage rooms	—	—	0.12	—
Warehouses (see storage)	—	—	—	—

(continued)

TABLE 403.3.1.1—continued
MINIMUM VENTILATION RATES

OCCUPANCY CLASSIFICATION	OCCUPANT DENSITY #/1000 FT² ª	PEOPLE OUTDOOR AIRFLOW RATE IN BREATHING ZONE, R_p CFM/PERSON	AREA OUTDOOR AIRFLOW RATE IN BREATHING ZONE, R_a CFM/FT² ª	EXHAUST AIRFLOW RATE CFM/FT² ª
Specialty shops				
Automotive motor-fuel dispensing stations[b]	—	—	—	1.5
Barber	25	7.5	0.06	0.5
Beauty salons[b]	25	20	0.12	0.6
Nail salons [b, h]	25	20	0.12	0.6
Embalming room[b]	—	—	—	2.0
Pet shops (animal areas)[b]	10	7.5	0.18	0.9
Supermarkets	8	7.5	0.06	—
Sports and amusement				
Bowling alleys (seating areas)	40	10	0.12	—
Disco/dance floors	100	20	0.06	—
Game arcades	20	7.5	0.18	—
Gym, stadium, arena (play area)	—	—	0.30	—
Health club/aerobics room	40	20	0.06	—
Health club/weight room	10	20	0.06	—
Ice arenas without combustion engines	—	—	0.30	0.5
Spectator areas	150	7.5	0.06	—
Swimming pools (pool and deck area)	—	—	0.48	—
Storage				
Repair garages, enclosed parking garages[b,d]	—	—	—	0.75
Warehouses	—	—	0.06	—
Theaters				
Auditoriums (see education)	—	—	—	—
Lobbies	150	5	0.06	—
Stages, studios	70	10	0.06	—
Ticket booths	60	5	0.06	—
Transportation				
Platforms	100	7.5	0.06	—
Transportation waiting	100	7.5	0.06	—

(continued)

TABLE 403.3.1.1—continued
MINIMUM VENTILATION RATES

OCCUPANCY CLASSIFICATION	OCCUPANT DENSITY #/1000 FT$^{2\,a}$	PEOPLE OUTDOOR AIRFLOW RATE IN BREATHING ZONE, R_p CFM/PERSON	AREA OUTDOOR AIRFLOW RATE IN BREATHING ZONE, R_a CFM/FT$^{2\,a}$	EXHAUST AIRFLOW RATE CFM/FT$^{2\,a}$
Workrooms				
Bank vaults/safe deposit	5	5	0.06	—
Computer (without printing)	4	5	0.06	—
Copy, printing rooms	4	5	0.06	0.5
Darkrooms	—	—	—	1.0
Meat processing^c	10	15	—	—
Pharmacy (prep. area)	10	5	0.18	—
Photo studios	10	5	0.12	—

For SI: 1 cubic foot per minute = 0.0004719 m^3/s, 1 ton = 908 kg, 1 cubic foot per minute per square foot = 0.00508 m^3/(s · m^2),
°C = [(°F) -32]/1.8, 1 square foot = 0.0929 m^2.

a. Based upon *net occupiable floor area.*

b. Mechanical exhaust required and the recirculation of air from such spaces is prohibited. Recirculation of air that is contained completely within such spaces shall not be prohibited (see Section 403.2.1, Item 3).

c. Spaces unheated or maintained below 50°F are not covered by these requirements unless the occupancy is continuous.

d. Ventilation systems in enclosed parking garages shall comply with Section 404.

e. Rates are per water closet or urinal. The higher rate shall be provided where the exhaust system is designed to operate intermittently. The lower rate shall be permitted only where the exhaust system is designed to operate continuously while occupied.

f. Rates are per room unless otherwise indicated. The higher rate shall be provided where the exhaust system is designed to operate intermittently. The lower rate shall be permitted only where the exhaust system is designed to operate continuously while occupied.

g. Mechanical exhaust is required and recirculation from such spaces is prohibited except that recirculation shall be permitted where the resulting supply airstream consists of not more than 10 percent air recirculated from these spaces. Recirculation of air that is contained completely within such spaces shall not be prohibited (see Section 403.2.1, Items 2 and 4).

h. For nail salons, each manicure and pedicure station shall be provided with a *source capture system* capable of exhausting not less than 50 cfm per station. Exhaust inlets shall be located in accordance with Section 502.20. Where one or more required source capture systems operate continuously during occupancy, the exhaust rate from such systems shall be permitted to be applied to the exhaust flow rate required by Table 403.3.1.1 for the nail salon.

CHAPTER 5

EXHAUST SYSTEMS

SECTION 501
GENERAL

501.1 Scope. This chapter shall govern the design, construction and installation of mechanical exhaust systems, including exhaust systems serving clothes dryers and cooking appliances; hazardous exhaust systems; dust, stock and refuse conveyor systems; subslab soil exhaust systems; smoke control systems; energy recovery ventilation systems and other systems specified in Section 502.

501.2 Independent system required. Single or combined mechanical exhaust systems for environmental air shall be independent of all other exhaust systems. Dryer exhaust shall be independent of all other systems. Type I exhaust systems shall be independent of all other exhaust systems except as provided in Section 506.3.5. Single or combined Type II exhaust systems for food-processing operations shall be independent of all other exhaust systems. Kitchen exhaust systems shall be constructed in accordance with Section 505 for domestic equipment and Sections 506 through 509 for commercial equipment.

501.3 Exhaust discharge. The air removed by every mechanical exhaust system shall be discharged outdoors at a point where it will not cause a public nuisance and not less than the distances specified in Section 501.3.1. The air shall be discharged to a location from which it cannot again be readily drawn in by a ventilating system. Air shall not be exhausted into an attic, crawl space, or be directed onto walkways.

Exceptions:

1. Whole-house ventilation-type attic fans shall be permitted to discharge into the attic space of *dwelling units* having private attics.

2. Commercial cooking recirculating systems.

3. Where installed in accordance with the manufacturer's instructions and where mechanical or *natural ventilation* is otherwise provided in accordance with Chapter 4, *listed* and *labeled* domestic ductless range hoods shall not be required to discharge to the outdoors.

501.3.1 Location of exhaust outlets. The termination point of exhaust outlets and ducts discharging to the outdoors shall be located with the following minimum distances:

1. For ducts conveying explosive or flammable vapors, fumes or dusts: 30 feet (9144 mm) from property lines; 10 feet (3048 mm) from operable openings into buildings; 6 feet (1829 mm) from exterior walls and roofs; 30 feet (9144 mm) from combustible walls and operable openings into buildings which are in the direction of the exhaust discharge; 10 feet (3048 mm) above adjoining grade.

2. For other product-conveying outlets: 10 feet (3048 mm) from the property lines; 3 feet (914 mm) from exterior walls and roofs; 10 feet (3048 mm) from operable openings into buildings; 10 feet (3048 mm) above adjoining grade.

3. For all *environmental air* exhaust: 3 feet (914 mm) from property lines; 3 feet (914 mm) from operable openings into buildings for all occupancies other than Group U, and 10 feet (3048 mm) from mechanical air intakes. Such exhaust shall not be considered hazardous or noxious.

4. Exhaust outlets serving structures in flood hazard areas shall be installed at or above the elevation required by Section 1612 of the *International Building Code* for utilities and attendant equipment.

5. For specific systems see the following sections:

   5.1. Clothes dryer exhaust, Section 504.4.

   5.2. Kitchen hoods and other kitchen exhaust *equipment*, Sections 506.3.13, 506.4 and 506.5.

   5.3. Dust stock and refuse conveying systems, Section 511.2.

   5.4. Subslab soil exhaust systems, Section 512.4.

   5.5. Smoke control systems, Section 513.10.3.

   5.6. Refrigerant discharge, Section 1105.7.

   5.7. Machinery room discharge, Section 1105.6.1.

501.3.2 Exhaust opening protection. Exhaust openings that terminate outdoors shall be protected with corrosion-resistant screens, louvers or grilles. Openings in screens, louvers and grilles shall be sized not less than $^1/_4$ inch (6.4 mm) and not larger than $^1/_2$ inch (12.7 mm). Openings shall be protected against local weather conditions. Louvers that protect exhaust openings in structures located in hurricane-prone regions, as defined in the *International Building Code*, shall comply with AMCA Standard 550. Outdoor openings located in exterior walls shall meet the provisions for exterior wall opening protectives in accordance with the *International Building Code*.

501.4 Pressure equalization. Mechanical exhaust systems shall be sized to remove the quantity of air required by this chapter to be exhausted. The system shall operate when air is required to be exhausted. Where mechanical exhaust is required in a room or space in other than occupancies in R-3 and *dwelling units* in R-2, such space shall be maintained with a neutral or negative pressure. If a greater quantity of air is supplied by a mechanical ventilating supply system than is removed by a mechanical exhaust for a room, adequate means shall be provided for the natural or mechanical exhaust of the excess air supplied. If only a mechanical exhaust system is installed for a room or if a greater quantity of air is removed by a mechanical exhaust system than is supplied by a mechanical ventilating supply system for a room, adequate *makeup air* shall be provided to satisfy the deficiency.

501.5 Ducts. Where exhaust duct construction is not specified in this chapter, such construction shall comply with Chapter 6.

SECTION 502
REQUIRED SYSTEMS

502.1 General. An exhaust system shall be provided, maintained and operated as specifically required by this section and for all occupied areas where machines, vats, tanks, furnaces, forges, salamanders and other *appliances, equipment* and processes in such areas produce or throw off dust or particles sufficiently light to float in the air, or which emit heat, odors, fumes, spray, gas or smoke, in such quantities so as to be irritating or injurious to health or safety.

502.1.1 Exhaust location. The inlet to an exhaust system shall be located in the area of heaviest concentration of contaminants.

[F] 502.1.2 Fuel-dispensing areas. The bottom of an air inlet or exhaust opening in fuel-dispensing areas shall be located not more than 18 inches (457 mm) above the floor.

502.1.3 Equipment, appliance and service rooms. *Equipment*, *appliance* and system service rooms that house sources of odors, fumes, noxious gases, smoke, steam, dust, spray or other contaminants shall be designed and constructed so as to prevent spreading of such contaminants to other occupied parts of the building.

[F] 502.1.4 Hazardous exhaust. The mechanical exhaust of high concentrations of dust or hazardous vapors shall conform to the requirements of Section 510.

[F] 502.2 Aircraft fueling and defueling. Compartments housing piping, pumps, air eliminators, water separators, hose reels and similar *equipment* used in aircraft fueling and defueling operations shall be adequately ventilated at floor level or within the floor itself.

[F] 502.3 Battery-charging areas for powered industrial trucks and equipment. Ventilation shall be provided in an *approved* manner in battery-charging areas for powered industrial trucks and *equipment* to prevent a dangerous accumulation of flammable gases.

[F] 502.4 Stationary storage battery systems. Stationary storage battery systems, as regulated by Section 608 of the *International Fire Code*, shall be provided with ventilation in accordance with this chapter and Section 502.4.1 or 502.4.2.

> **Exception:** Lithium-ion and lithium metal polymer batteries shall not require additional ventilation beyond that which would normally be required for human occupancy of the space.

[F] 502.4.1 Hydrogen limit in rooms. For flooded lead acid, flooded nickel cadmium and VRLA batteries, the ventilation system shall be designed to limit the maximum concentration of hydrogen to 1.0 percent of the total volume of the room.

[F] 502.4.2 Ventilation rate in rooms. Continuous ventilation shall be provided at a rate of not less than 1 cubic foot per minute per square foot (cfm/ft^2) [0.00508 m^3/(s • m^2)] of floor area of the room.

[F] 502.4.3 Supervision. Mechanical ventilation systems required by Section 502.4 shall be supervised by an approved central, proprietary or remote station service or shall initiate an audible and visual signal at a constantly attended on-site location.

[F] 502.5 Valve-regulated lead-acid batteries in cabinets. Valve-regulated lead-acid (VRLA) batteries installed in cabinets, as regulated by Section 608.6.2 of the *International Fire Code*, shall be provided with ventilation in accordance with Section 502.5.1 or 502.5.2.

[F] 502.5.1 Hydrogen limit in cabinets. The cabinet ventilation system shall be designed to limit the maximum concentration of hydrogen to 1.0 percent of the total volume of the cabinet during the worst-case event of simultaneous boost charging of all batteries in the cabinet.

[F] 502.5.2 Ventilation rate in cabinets. Continuous cabinet ventilation shall be provided at a rate of not less than 1 cubic foot per minute per square foot (cfm/ft^2) [0.00508 m^3/(s • m^2)] of the floor area covered by the cabinet. The room in which the cabinet is installed shall be ventilated as required by Section 502.4.1 or 502.4.2.

[F] 502.5.3 Supervision. Mechanical ventilation systems required by Section 502.5 shall be supervised by an approved central, proprietary or remote station service or shall initiate an audible and visual signal at a constantly attended on-site location.

[F] 502.6 Dry cleaning plants. Ventilation in dry cleaning plants shall be adequate to protect employees and the public in accordance with this section and DOL 29 CFR Part 1910.1000, where applicable.

[F] 502.6.1 Type II systems. Type II dry cleaning systems shall be provided with a mechanical ventilation system that is designed to exhaust 1 cubic foot of air per minute for each square foot of floor area (1 cfm/ft^2) [0.00508 m^3/ (s • m^2)] in dry cleaning rooms and in drying rooms. The ventilation system shall operate automatically when the dry cleaning *equipment* is in operation and shall have manual controls at an *approved* location.

[F] 502.6.2 Type IV and V systems. Type IV and V dry cleaning systems shall be provided with an automatically activated exhaust ventilation system to maintain an air velocity of not less than 100 feet per minute (0.51 m/s) through the loading door when the door is opened.

> **Exception:** Dry cleaning units are not required to be provided with exhaust ventilation where an exhaust hood is installed immediately outside of and above the loading door which operates at an airflow rate as follows:

$$Q = 100 \times A_{LD} \qquad \text{Equation 5-1)}$$

where:

Q = Flow rate exhausted through the hood, cubic feet per minute.

A_{LD} = Area of the loading door, square feet.

[F] 502.6.3 Spotting and pretreating. Scrubbing tubs, scouring, brushing or spotting operations shall be located such that solvent vapors are captured and exhausted by the ventilating system.

[F] 502.7 Application of flammable finishes. Mechanical exhaust as required by this section shall be provided for operations involving the application of flammable finishes.

[F] 502.7.1 During construction. Ventilation shall be provided for operations involving the application of materials containing flammable solvents in the course of construction, *alteration* or demolition of a structure.

[F] 502.7.2 Limited spraying spaces. Positive mechanical ventilation that provides not less than six complete air changes per hour shall be installed in limited spraying spaces. Such system shall meet the requirements of the *International Fire Code* for handling flammable vapors. Explosion venting is not required.

[F] 502.7.3 Flammable vapor areas. Mechanical ventilation of flammable vapor areas shall be provided in accordance with Sections 502.7.3.1 through 502.7.3.6.

[F] 502.7.3.1 Operation. Mechanical ventilation shall be kept in operation at all times while spraying operations are being conducted and for a sufficient time thereafter to allow vapors from drying coated articles and finishing material residue to be exhausted. Spraying *equipment* shall be interlocked with the ventilation of the flammable vapor area such that spraying operations cannot be conducted unless the ventilation system is in operation.

[F] 502.7.3.2 Recirculation. Air exhausted from spraying operations shall not be recirculated.

Exceptions:

1. Air exhausted from spraying operations shall be permitted to be recirculated as *makeup air* for unmanned spray operations provided that:

   1.1. The solid particulate has been removed.

   1.2. The vapor concentration is less than 25 percent of the lower flammable limit (LFL).

   1.3. *Approved equipment* is used to monitor the vapor concentration.

   1.4. An alarm is sounded and spray operations are automatically shut down if the vapor concentration exceeds 25 percent of the LFL.

   1.5. In the event of shutdown of the vapor concentration monitor, 100 percent of the air volume specified in Section 510 is automatically exhausted.

2. Air exhausted from spraying operations is allowed to be recirculated as *makeup air* to manned spraying operations where all of the conditions provided in Exception 1 are included in the installation and documents have been prepared to show that the installation does not pose a life safety hazard to personnel inside the spray booth, spraying space or spray room.

[F] 502.7.3.3 Air velocity. The ventilation system shall be designed, installed and maintained so that the flammable contaminants are diluted in noncontaminated air to maintain concentrations in the exhaust air flow below 25 percent of the contaminant's lower flammable limit (LFL). In addition, the spray booth shall be provided with mechanical ventilation so that the average air velocity through openings is in accordance with Sections 502.7.3.3.1 and 502.7.3.3.2.

[F] 502.7.3.3.1 Open face or open front spray booth. For spray application operations conducted in an open face or open front spray booth, the ventilation system shall be designed, installed and maintained so that the average air velocity into the spray booth through all openings is not less than 100 feet per minute (0.51 m/s).

Exception: For fixed or automated electrostatic spray application equipment, the average air velocity into the spray booth through all openings shall be not less than 50 feet per minute (0.25 m/s).

[F] 502.7.3.3.2 Enclosed spray booth or spray room with openings for product conveyance. For spray application operations conducted in an enclosed spray booth or spray room with openings for product conveyance, the ventilation system shall be designed, installed and maintained so that the average air velocity into the spray booth through openings is not less than 100 feet per minute (0.51 m/s).

Exceptions:

1. For fixed or automated electrostatic spray application equipment, the average air velocity into the spray booth through all openings shall be not less than 50 feet per minute (0.25 m/s).

2. Where methods are used to reduce cross drafts that can draw vapors and overspray through openings from the spray booth or spray room, the average air velocity into the spray booth or spray room shall be that necessary to capture and confine vapors and overspray to the spray booth or spray room.

[F] 502.7.3.4 Ventilation obstruction. Articles being sprayed shall be positioned in a manner that does not obstruct collection of overspray.

[F] 502.7.3.5 Independent ducts. Each spray booth and spray room shall have an independent exhaust duct system discharging to the outdoors.

Exceptions:

1. Multiple spray booths having a combined frontal area of 18 square feet (1.67 m²) or less are allowed to have a common exhaust where identical spray-finishing material is used in each booth. If more than one fan serves one booth, such fans shall be interconnected so that all fans operate simultaneously.

2. Where treatment of exhaust is necessary for air pollution control or energy conservation, ducts shall be allowed to be manifolded if all of the following conditions are met:

   2.1. The sprayed materials used are compatible and will not react or cause ignition of the residue in the ducts.

   2.2. Nitrocellulose-based finishing material shall not be used.

   2.3. A filtering system shall be provided to reduce the amount of overspray carried into the duct manifold.

   2.4. Automatic sprinkler protection shall be provided at the junction of each booth exhaust with the manifold, in addition to the protection required by this chapter.

[F] 502.7.3.6 Fan motors and belts. Electric motors driving exhaust fans shall not be placed inside booths or ducts. Fan rotating elements shall be nonferrous or nonsparking or the casing shall consist of, or be lined with, such material. Belts shall not enter the duct or booth unless the belt and pulley within the duct are tightly enclosed.

[F] 502.7.4 Dipping operations. Flammable vapor areas of dip tank operations shall be provided with mechanical ventilation adequate to prevent the dangerous accumulation of vapors. Required ventilation systems shall be so arranged that the failure of any ventilating fan will automatically stop the dipping conveyor system.

[F] 502.7.5 Electrostatic apparatus. The flammable vapor area in spray-finishing operations involving electrostatic apparatus and devices shall be ventilated in accordance with Section 502.7.3.

[F] 502.7.6 Powder coating. Exhaust ventilation for powder-coating operations shall be sufficient to maintain the atmosphere below one-half of the minimum explosive concentration for the material being applied. Nondeposited, air-suspended powders shall be removed through exhaust ducts to the powder recovery system.

[F] 502.7.7 Floor resurfacing operations. To prevent the accumulation of flammable vapors during floor resurfacing operations, mechanical ventilation at a minimum rate of 1 cfm/ft² [0.00508 m³/(s • m²)] of area being finished shall be provided. Such exhaust shall be by *approved* temporary or portable means. Vapors shall be exhausted to the exterior of the building.

[F] 502.8 Hazardous materials—general requirements. Exhaust ventilation systems for structures containing hazardous materials shall be provided as required in Sections 502.8.1 through 502.8.5.

[F] 502.8.1 Storage in excess of the maximum allowable quantities. Indoor storage areas and storage buildings for hazardous materials in amounts exceeding the maximum allowable quantity per control area shall be provided with mechanical exhaust ventilation or *natural ventilation* where *natural ventilation* can be shown to be acceptable for the materials as stored.

Exceptions:

1. Storage areas for flammable solids complying with Section 5904 of the *International Fire Code*.

2. Storage areas and storage buildings for fireworks and explosives complying with Chapter 56 of the *International Fire Code*.

[F] 502.8.1.1 System requirements. Exhaust ventilation systems shall comply with all of the following:

1. The installation shall be in accordance with this code.

2. Mechanical ventilation shall be provided at a rate of not less than 1 cfm per square foot [0.00508 m³/(s • m²)] of floor area over the storage area.

3. The systems shall operate continuously unless alternate designs are *approved*.

4. A manual shutoff control shall be provided outside of the room in a position adjacent to the access door to the room or in another *approved* location. The switch shall be a break-glass or other *approved* type and shall be *labeled*: VENTILATION SYSTEM EMERGENCY SHUT-OFF.

5. The exhaust ventilation shall be designed to consider the density of the potential fumes or vapors released. For fumes or vapors that are heavier than air, exhaust shall be taken from a point within 12 inches (305 mm) of the floor. For fumes or vapors that are lighter than air, exhaust shall be taken from a point within 12 inches (305 mm) of the highest point of the room.

6. The location of both the exhaust and inlet air openings shall be designed to provide air movement across all portions of the floor or room to prevent the accumulation of vapors.

7. The *exhaust air* shall not be recirculated to occupied areas if the materials stored are capable of emitting hazardous vapors and contaminants

have not been removed. Air contaminated with explosive or flammable vapors, fumes or dusts; flammable, highly toxic or toxic gases; or radioactive materials shall not be recirculated.

[F] 502.8.2 Gas rooms, exhausted enclosures and gas cabinets. The ventilation system for gas rooms, exhausted enclosures and gas cabinets for any quantity of hazardous material shall be designed to operate at a negative pressure in relation to the surrounding area. Highly toxic and toxic gases shall comply with Sections 502.9.7.1, 502.9.7.2 and 502.9.8.4.

[F] 502.8.3 Indoor dispensing and use. Indoor dispensing and use areas for hazardous materials in amounts exceeding the maximum allowable quantity per control area shall be provided with exhaust ventilation in accordance with Section 502.8.1.

> **Exception:** Ventilation is not required for dispensing and use of flammable solids other than finely divided particles.

[F] 502.8.4 Indoor dispensing and use—point sources. Where gases, liquids or solids in amounts exceeding the maximum allowable quantity per control area and having a hazard ranking of 3 or 4 in accordance with NFPA 704 are dispensed or used, mechanical exhaust ventilation shall be provided to capture gases, fumes, mists or vapors at the point of generation.

> **Exception:** Where it can be demonstrated that the gases, liquids or solids do not create harmful gases, fumes, mists or vapors.

[F] 502.8.5 Closed systems. Where closed systems for the use of hazardous materials in amounts exceeding the maximum allowable quantity per control area are designed to be opened as part of normal operations, ventilation shall be provided in accordance with Section 502.8.4.

[F] 502.9 Hazardous materials—requirements for specific materials. Exhaust ventilation systems for specific hazardous materials shall be provided as required in Section 502.8 and Sections 502.9.1 through 502.9.11.

[F] 502.9.1 Compressed gases—medical gas systems. Rooms for the storage of compressed medical gases in amounts exceeding the permit amounts for compressed gases in the *International Fire Code*, and that do not have an exterior wall, shall be exhausted through a duct to the exterior of the building. Both separate airstreams shall be enclosed in a 1-hour-rated shaft enclosure from the room to the exterior. *Approved* mechanical ventilation shall be provided at a minimum rate of 1 cfm/ft² [0.00508 m³/(s • m²)] of the area of the room.

Gas cabinets for the storage of compressed medical gases in amounts exceeding the permit amounts for compressed gases in the *International Fire Code* shall be connected to an exhaust system. The average velocity of ventilation at the face of access ports or windows shall be not less than 200 feet per minute (1.02 m/s) with a minimum velocity of 150 feet per minute (0.76 m/s) at any point at the access port or window.

[F] 502.9.2 Corrosives. Where corrosive materials in amounts exceeding the maximum allowable quantity per control area are dispensed or used, mechanical exhaust ventilation in accordance with Section 502.8.4 shall be provided.

[F] 502.9.3 Cryogenics. Storage areas for stationary or portable containers of cryogenic fluids in any quantity shall be ventilated in accordance with Section 502.8. Indoor areas where cryogenic fluids in any quantity are dispensed shall be ventilated in accordance with the requirements of Section 502.8.4 in a manner that captures any vapor at the point of generation.

> **Exception:** Ventilation for indoor dispensing areas is not required where it can be demonstrated that the cryogenic fluids do not create harmful vapors.

[F] 502.9.4 Explosives. Squirrel cage blowers shall not be used for exhausting hazardous fumes, vapors or gases in operating buildings and rooms for the manufacture, assembly or testing of explosives. Only nonferrous fan blades shall be used for fans located within the ductwork and through which hazardous materials are exhausted. Motors shall be located outside the duct.

[F] 502.9.5 Flammable and combustible liquids. Exhaust ventilation systems shall be provided as required by Sections 502.9.5.1 through 502.9.5.5 for the storage, use, dispensing, mixing and handling of flammable and combustible liquids. Unless otherwise specified, this section shall apply to any quantity of flammable and combustible liquids.

> **Exception:** This section shall not apply to flammable and combustible liquids that are exempt from the *International Fire Code*.

[F] 502.9.5.1 Vaults. Vaults that contain tanks of Class I liquids shall be provided with continuous ventilation at a rate of not less than 1 cfm/ft² of floor area [0.00508 m³/(s • m²)], but not less than 150 cfm (4 m³/min). Failure of the exhaust airflow shall automatically shut down the dispensing system. The exhaust system shall be designed to provide air movement across all parts of the vault floor. Supply and exhaust ducts shall extend to a point not greater than 12 inches (305 mm) and not less than 3 inches (76 mm) above the floor. The exhaust system shall be installed in accordance with the provisions of NFPA 91. Means shall be provided to automatically detect any flammable vapors and to automatically shut down the dispensing system upon detection of such flammable vapors in the exhaust duct at a concentration of 25 percent of the LFL.

[F] 502.9.5.2 Storage rooms and warehouses. Liquid storage rooms and liquid storage warehouses for quantities of liquids exceeding those specified in the *International Fire Code* shall be ventilated in accordance with Section 502.8.1.

[F] 502.9.5.3 Cleaning machines. Areas containing machines used for parts cleaning in accordance with the *International Fire Code* shall be adequately ventilated to prevent accumulation of vapors.

[F] 502.9.5.4 Use, dispensing and mixing. Continuous mechanical ventilation shall be provided for the use, dispensing and mixing of flammable and combustible liquids in open or closed systems in amounts exceeding the maximum allowable quantity per control area and for bulk transfer and process transfer operations. The ventilation rate shall be not less than 1 cfm/ft² [0.00508 m³/(s • m²)] of floor area over the design area. Provisions shall be made for the introduction of *makeup air* in a manner that will include all floor areas or pits where vapors can collect. Local or spot ventilation shall be provided where needed to prevent the accumulation of hazardous vapors.

> **Exception:** Where *natural ventilation* can be shown to be effective for the materials used, dispensed or mixed.

[F] 502.9.5.5 Bulk plants or terminals. Ventilation shall be provided for portions of properties where flammable and combustible liquids are received by tank vessels, pipelines, tank cars or tank vehicles and which are stored or blended in bulk for the purpose of distributing such liquids by tank vessels, pipelines, tank cars, tank vehicles or containers as required by Sections 502.9.5.5.1 through 502.9.5.5.3.

[F] 502.9.5.5.1 General. Ventilation shall be provided for rooms, buildings and enclosures in which Class I liquids are pumped, used or transferred. Design of ventilation systems shall consider the relatively high specific gravity of the vapors. Where *natural ventilation* is used, adequate openings in outside walls at floor level, unobstructed except by louvers or coarse screens, shall be provided. Where *natural ventilation* is inadequate, mechanical ventilation shall be provided.

[F] 502.9.5.5.2 Basements and pits. Class I liquids shall not be stored or used within a building having a basement or pit into which flammable vapors can travel, unless such area is provided with ventilation designed to prevent the accumulation of flammable vapors therein.

[F] 502.9.5.5.3 Dispensing of Class I liquids. Containers of Class I liquids shall not be drawn from or filled within buildings unless a provision is made to prevent the accumulation of flammable vapors in hazardous concentrations. Where mechanical ventilation is required, it shall be kept in operation while flammable vapors could be present.

[F] 502.9.6 Highly toxic and toxic liquids. Ventilation exhaust shall be provided for highly toxic and toxic liquids as required by Sections 502.9.6.1 and 502.9.6.2.

[F] 502.9.6.1 Treatment system. This provision shall apply to indoor and outdoor storage and use of highly toxic and toxic liquids in amounts exceeding the maximum allowable quantities per control area. Exhaust scrubbers or other systems for processing vapors of highly toxic liquids shall be provided where a spill or accidental release of such liquids can be expected to release highly toxic vapors at normal temperature and pressure.

[F] 502.9.6.2 Open and closed systems. Mechanical exhaust ventilation shall be provided for highly toxic and toxic liquids used in open systems in accordance with Section 502.8.4. Mechanical exhaust ventilation shall be provided for highly toxic and toxic liquids used in closed systems in accordance with Section 502.8.5.

> **Exception:** Liquids or solids that do not generate highly toxic or toxic fumes, mists or vapors.

[F] 502.9.7 Highly toxic and toxic compressed gases—any quantity. Ventilation exhaust shall be provided for highly toxic and toxic compressed gases in any quantity as required by Sections 502.9.7.1 and 502.9.7.2.

[F] 502.9.7.1 Gas cabinets. Gas cabinets containing highly toxic or toxic compressed gases in any quantity shall comply with Section 502.8.2 and the following requirements:

1. The average ventilation velocity at the face of gas cabinet access ports or windows shall be not less than 200 feet per minute (1.02 m/s) with a minimum velocity of 150 feet per minute (0.76 m/s) at any point at the access port or window.

2. Gas cabinets shall be connected to an exhaust system.

3. Gas cabinets shall not be used as the sole means of exhaust for any room or area.

[F] 502.9.7.2 Exhausted enclosures. Exhausted enclosures containing highly toxic or toxic compressed gases in any quantity shall comply with Section 502.8.2 and the following requirements:

1. The average ventilation velocity at the face of the enclosure shall be not less than 200 feet per minute (1.02 m/s) with a minimum velocity of 150 feet per minute (0.76 m/s).

2. Exhausted enclosures shall be connected to an exhaust system.

3. Exhausted enclosures shall not be used as the sole means of exhaust for any room or area.

[F] 502.9.8 Highly toxic and toxic compressed gases—quantities exceeding the maximum allowable quantity per control area. Ventilation exhaust shall be provided for highly toxic and toxic compressed gases in amounts exceeding the maximum allowable quantities per control area as required by Sections 502.9.8.1 through 502.9.8.6.

[F] 502.9.8.1 Ventilated areas. The room or area in which indoor gas cabinets or exhausted enclosures are located shall be provided with exhaust ventilation. Gas cabinets or exhausted enclosures shall not be used as the sole means of exhaust for any room or area.

[F] 502.9.8.2 Local exhaust for portable tanks. A means of local exhaust shall be provided to capture leakage from indoor and outdoor portable tanks. The local exhaust shall consist of portable ducts or collection systems designed to be applied to the site of a leak in a valve or fitting on the tank. The local exhaust system shall be located in a gas room. Exhaust shall be directed to a treatment system where required by the *International Fire Code*.

[F] 502.9.8.3 Piping and controls—stationary tanks. Filling or dispensing connections on indoor stationary tanks shall be provided with a means of local exhaust. Such exhaust shall be designed to capture fumes and vapors. The exhaust shall be directed to a treatment system where required by the *International Fire Code*.

[F] 502.9.8.4 Gas rooms. The ventilation system for gas rooms shall be designed to operate at a negative pressure in relation to the surrounding area. The exhaust ventilation from gas rooms shall be directed to an exhaust system.

[F] 502.9.8.5 Treatment system. The exhaust ventilation from gas cabinets, exhausted enclosures and gas rooms, and local exhaust systems required in Sections 502.9.8.2 and 502.9.8.3 shall be directed to a treatment system where required by the *International Fire Code*.

[F] 502.9.8.6 Process equipment. Effluent from indoor and outdoor process *equipment* containing highly toxic or toxic compressed gases which could be discharged to the atmosphere shall be processed through an exhaust scrubber or other processing system. Such systems shall be in accordance with the *International Fire Code*.

[F] 502.9.9 Ozone gas generators. Ozone cabinets and ozone gas-generator rooms for systems having a maximum ozone-generating capacity of $^1/_2$ pound (0.23 kg) or more over a 24-hour period shall be mechanically ventilated at a rate of not less than six air changes per hour. For cabinets, the average velocity of ventilation at *makeup air* openings with cabinet doors closed shall be not less than 200 feet per minute (1.02 m/s).

[F] 502.9.10 LP-gas distribution facilities. LP-gas distribution facilities shall be ventilated in accordance with NFPA 58.

[F] 502.9.10.1 Portable container use. Above-grade underfloor spaces or basements in which portable LP-gas containers are used or are stored awaiting use or resale shall be provided with an *approved* means of ventilation.

Exception: Department of Transportation (DOT) specification cylinders with a maximum water capacity of 2.5 pounds (1 kg) for use in completely self-contained hand torches and similar applications. The quantity of LP-gas shall not exceed 20 pounds (9 kg).

[F] 502.9.11 Silane gas. Exhausted enclosures and gas cabinets for the indoor storage of silane gas in amounts exceeding the maximum allowable quantities per control area shall comply with Chapter 64 of the *International Fire Code*.

[F] 502.10 Hazardous production materials (HPM). Exhaust ventilation systems and materials for ducts utilized for the exhaust of HPM shall comply with this section, other applicable provisions of this code, the *International Building Code* and the *International Fire Code*.

[F] 502.10.1 Where required. Exhaust ventilation systems shall be provided in the following locations in accordance with the requirements of this section and the *International Building Code*.

1. Fabrication areas: Exhaust ventilation for fabrication areas shall comply with the *International Building Code*. Additional manual control switches shall be provided where required by the code official.

2. Workstations: A ventilation system shall be provided to capture and exhaust gases, fumes and vapors at workstations.

3. Liquid storage rooms: Exhaust ventilation for liquid storage rooms shall comply with Section 502.8.1.1 and the *International Building Code*.

4. HPM rooms: Exhaust ventilation for HPM rooms shall comply with Section 502.8.1.1 and the *International Building Code*.

5. Gas cabinets: Exhaust ventilation for gas cabinets shall comply with Section 502.8.2. The gas cabinet ventilation system is allowed to connect to a workstation ventilation system. Exhaust ventilation for gas cabinets containing highly toxic or toxic gases shall also comply with Sections 502.9.7 and 502.9.8.

6. Exhausted enclosures: Exhaust ventilation for exhausted enclosures shall comply with Section 502.8.2. Exhaust ventilation for exhausted enclosures containing highly toxic or toxic gases shall also comply with Sections 502.9.7 and 502.9.8.

7. Gas rooms: Exhaust ventilation for gas rooms shall comply with Section 502.8.2. Exhaust ventilation for gas rooms containing highly toxic or toxic gases shall also comply with Sections 502.9.7 and 502.9.8.

8. Cabinets containing pyrophoric liquids or Class 3 water-reactive liquids: Exhaust ventilation for cabinets in fabrication areas containing pyrophoric liquids shall be as required in Section 2705.2.3.4 of the *International Fire Code*.

[F] 502.10.2 Penetrations. Exhaust ducts penetrating fire barriers constructed in accordance with Section 707 of the *International Building Code* or horizontal assemblies constructed in accordance with Section 711 of the *International Building Code* shall be contained in a shaft of equivalent fire-resistance-rated construction. Exhaust ducts shall not penetrate fire walls. Fire dampers shall not be installed in exhaust ducts.

[F] 502.10.3 Treatment systems. Treatment systems for highly toxic and toxic gases shall comply with the *International Fire Code*.

502.11 Motion picture projectors. Motion picture projectors shall be exhausted in accordance with Section 502.11.1 or 502.11.2.

502.11.1 Projectors with an exhaust discharge. Projectors equipped with an exhaust discharge shall be directly connected to a mechanical exhaust system. The exhaust system shall operate at an exhaust rate as indicated by the manufacturer's installation instructions.

502.11.2 Projectors without exhaust connection. Projectors without an exhaust connection shall have contaminants exhausted through a mechanical exhaust system. The exhaust rate for electric arc projectors shall be not less than 200 cubic feet per minute (cfm) (0.09 m^3/s) per lamp. The exhaust rate for xenon projectors shall be not less than 300 cfm (0.14 m^3/s) per lamp. Xenon projector exhaust shall be at a rate such that the exterior temperature of the lamp housing does not exceed 130°F (54°C). The lamp and projection room exhaust systems, whether combined or independent, shall not be interconnected with any other exhaust or return system within the building.

[F] 502.12 Organic coating processes. Enclosed structures involving organic coating processes in which Class I liquids are processed or handled shall be ventilated at a rate of not less than 1 cfm/ft^2 [0.00508 m^3/(s • m^2)] of solid floor area. Ventilation shall be accomplished by exhaust fans that intake at floor levels and discharge to a safe location outside the structure. Noncontaminated intake air shall be introduced in such a manner that all portions of solid floor areas are provided with continuous uniformly distributed air movement.

502.13 Public garages. Mechanical exhaust systems for public garages, as required in Chapter 4, shall operate continuously or in accordance with Section 404.

502.14 Motor vehicle operation. In areas where motor vehicles operate, mechanical ventilation shall be provided in accordance with Section 403. Additionally, areas in which stationary motor vehicles are operated shall be provided with a *source capture system* that connects directly to the motor vehicle exhaust systems. Such system shall be engineered by a registered design professional or shall be factory-built equipment designed and sized for the purpose.

Exceptions:

1. This section shall not apply where the motor vehicles being operated or repaired are electrically powered.

2. This section shall not apply to one- and two-family dwellings.

3. This section shall not apply to motor vehicle service areas where engines are operated inside the building only for the duration necessary to move the motor vehicles in and out of the building.

[F] 502.15 Repair garages. Where Class I liquids or LP-gas are stored or used within a building having a basement or pit wherein flammable vapors could accumulate, the basement or pit shall be provided with ventilation designed to prevent the accumulation of flammable vapors therein.

[F] 502.16 Repair garages for natural gas- and hydrogen-fueled vehicles. Repair garages used for the repair of natural gas- or hydrogen-fueled vehicles shall be provided with an *approved* mechanical ventilation system. The mechanical ventilation system shall be in accordance with Sections 502.16.1 and 502.16.2.

Exception: Where *approved* by the code official, *natural ventilation* shall be permitted in lieu of mechanical ventilation.

[F] 502.16.1 Design. Indoor locations shall be ventilated utilizing air supply inlets and exhaust outlets arranged to provide uniform air movement to the extent practical. Inlets shall be uniformly arranged on exterior walls near floor level. Outlets shall be located at the high point of the room in exterior walls or the roof.

Ventilation shall be by a continuous mechanical ventilation system or by a mechanical ventilation system activated by a continuously monitoring natural gas detection system, or for hydrogen, a continuously monitoring flammable gas detection system, each activating at a gas concentration of 25 percent of the lower flammable limit (LFL). In all cases, the system shall shut down the fueling system in the event of failure of the ventilation system.

The ventilation rate shall be not less than 1 cubic foot per minute per 12 cubic feet [0.00138 m^3/(s • m^3)] of room volume.

[F] 502.16.2 Operation. The mechanical ventilation system shall operate continuously.

Exceptions:

1. Mechanical ventilation systems that are interlocked with a gas detection system designed in accordance with the *International Fire Code*.

2. Mechanical ventilation systems in garages that are used only for the repair of vehicles fueled by liquid fuels or odorized gases, such as CNG, where the ventilation system is electrically interlocked with the lighting circuit.

502.17 Tire rebuilding or recapping. Each room where rubber cement is used or mixed, or where flammable or combustible solvents are applied, shall be ventilated in accordance with the applicable provisions of NFPA 91.

502.17.1 Buffing machines. Each buffing machine shall be connected to a dust-collecting system that prevents the accumulation of the dust produced by the buffing process.

502.18 Specific rooms. Specific rooms, including bathrooms, locker rooms, smoking lounges and toilet rooms, shall be exhausted in accordance with the ventilation requirements of Chapter 4.

502.19 Indoor firing ranges. Ventilation shall be provided in an *approved* manner in areas utilized as indoor firing ranges. Ventilation shall be designed to protect employees and the public in accordance with DOL 29 CFR 1910.1025 where applicable.

502.20 Manicure and pedicure stations. Manicure and pedicure stations shall be provided with an exhaust system in accordance with Table 403.3.1.1, Note h. Manicure tables and pedicure stations not provided with factory-installed exhaust inlets shall be provided with exhaust inlets located not more than 12 inches (305 mm) horizontally and vertically from the point of chemical application.

SECTION 503
MOTORS AND FANS

503.1 General. Motors and fans shall be sized to provide the required air movement. Motors in areas that contain flammable vapors or dusts shall be of a type *approved* for such environments. A manually operated remote control installed at an *approved* location shall be provided to shut off fans or blowers in flammable vapor or dust systems. Electrical *equipment* and appliances used in operations that generate explosive or flammable vapors, fumes or dusts shall be interlocked with the ventilation system so that the *equipment* and appliances cannot be operated unless the ventilation fans are in operation. Motors for fans used to convey flammable vapors or dusts shall be located outside the duct or shall be protected with *approved* shields and dustproofing. Motors and fans shall be provided with a means of access for servicing and maintenance.

503.2 Fans. Parts of fans in contact with explosive or flammable vapors, fumes or dusts shall be of nonferrous or nonsparking materials, or their casing shall be lined or constructed of such material. Where the size and hardness of materials passing through a fan are capable of producing a spark, both the fan and the casing shall be of nonsparking materials. Where fans are required to be spark resistant, their bearings shall not be within the airstream, and all parts of the fan shall be grounded. Fans in systems-handling materials that are capable of clogging the blades, and fans in buffing or woodworking exhaust systems, shall be of the radial-blade or tube-axial type.

503.3 Equipment and appliance identification plate. *Equipment* and appliances used to exhaust explosive or flammable vapors, fumes or dusts shall bear an identification plate stating the ventilation rate for which the system was designed.

503.4 Corrosion-resistant fans. Fans located in systems conveying corrosives shall be of materials that are resistant to the corrosive or shall be coated with corrosion-resistant materials.

SECTION 504
CLOTHES DRYER EXHAUST

504.1 Installation. Clothes dryers shall be exhausted in accordance with the manufacturer's instructions. Dryer exhaust systems shall be independent of all other systems and shall convey the moisture and any products of *combustion* to the outside of the building.

> **Exception:** This section shall not apply to *listed* and *labeled* condensing (ductless) clothes dryers.

504.2 Exhaust penetrations. Where a clothes dryer exhaust duct penetrates a wall or ceiling membrane, the annular space shall be sealed with noncombustible material, *approved* fire caulking or a noncombustible dryer exhaust duct wall receptacle. Ducts that exhaust clothes dryers shall not penetrate or be located within any fireblocking, draftstopping or any wall, floor/ceiling or other assembly required by the *International Building Code* to be fire-resistance rated, unless such duct is constructed of galvanized steel or aluminum of the thickness specified in Section 603.4 and the fire-resistance rating is maintained in accordance with the *International Building Code*. Fire dampers, combination fire/smoke dampers and any similar devices that will obstruct the exhaust flow shall be prohibited in clothes dryer exhaust ducts.

504.3 Cleanout. Each vertical riser shall be provided with a means for cleanout.

504.4 Exhaust installation. Dryer exhaust ducts for clothes dryers shall terminate on the outside of the building and shall be equipped with a backdraft damper. Screens shall not be installed at the duct termination. Ducts shall not be connected or installed with sheet metal screws or other fasteners that will obstruct the exhaust flow. Clothes dryer exhaust ducts shall not be connected to a vent connector, vent or *chimney*. Clothes dryer exhaust ducts shall not extend into or through ducts or plenums.

504.5 Dryer exhaust duct power ventilators. Domestic dryer exhaust duct power ventilators shall be listed and labeled to UL 705 for use in dryer exhaust duct systems. The dryer exhaust duct power ventilator shall be installed in accordance with the manufacturer's instructions.

504.6 Makeup air. Installations exhausting more than 200 cfm (0.09 m^3/s) shall be provided with *makeup air*. Where a closet is designed for the installation of a clothes dryer, an opening having an area of not less than 100 square inches (0.0645 m^2) shall be provided in the closet enclosure or *makeup air* shall be provided by other *approved* means.

504.7 Protection required. Protective shield plates shall be placed where nails or screws from finish or other work are likely to penetrate the clothes dryer exhaust duct. Shield plates shall be placed on the finished face of all framing members where there is less than $1^1/_4$ inches (32 mm) between the duct and the finished face of the framing member. Protective shield plates shall be constructed of steel, have a thickness of 0.062 inch (1.6 mm) and extend not less than 2 inches (51 mm) above sole plates and below top plates.

504.8 Domestic clothes dryer ducts. Exhaust ducts for domestic clothes dryers shall conform to the requirements of Sections 504.8.1 through 504.8.6.

> **504.8.1 Material and size.** Exhaust ducts shall have a smooth interior finish and shall be constructed of metal a minimum 0.016 inch (0.4 mm) thick. The exhaust duct size shall be 4 inches (102 mm) nominal in diameter.

> **504.8.2 Duct installation.** Exhaust ducts shall be supported at 4-foot (1219 mm) intervals and secured in place. The insert end of the duct shall extend into the adjoining duct or fitting in the direction of airflow. Ducts shall not be joined with screws or similar fasteners that protrude more than $1/_8$ inch (3.2 mm) into the inside of the duct.

> **504.8.3 Transition ducts.** Transition ducts used to connect the dryer to the exhaust duct system shall be a single length that is *listed* and *labeled* in accordance with UL 2158A. Transition ducts shall be not greater than 8 feet (2438 mm) in length and shall not be concealed within construction.

504.8.4 Duct length. The maximum allowable exhaust duct length shall be determined by one of the methods specified in Sections 504.8.4.1 through 504.8.4.3.

504.8.4.1 Specified length. The maximum length of the exhaust duct shall be 35 feet (10 668 mm) from the connection to the transition duct from the dryer to the outlet terminal. Where fittings are used, the maximum length of the exhaust duct shall be reduced in accordance with Table 504.8.4.1.

TABLE 504.8.4.1
DRYER EXHAUST DUCT FITTING EQUIVALENT LENGTH

DRYER EXHAUST DUCT FITTING TYPE	EQUIVALENT LENGTH
4″ radius mitered 45-degree elbow	2 feet 6 inches
4″ radius mitered 90-degree elbow	5 feet
6″ radius smooth 45-degree elbow	1 foot
6″ radius smooth 90-degree elbow	1 foot 9 inches
8″ radius smooth 45-degree elbow	1 foot
8″ radius smooth 90-degree elbow	1 foot 7 inches
10″ radius smooth 45-degree elbow	9 inches
10″ radius smooth 90-degree elbow	1 foot 6 inches

For SI: 1 inch = 25.4 mm, 1 foot = 304.8 mm, 1 degree = 0.0175 rad.

504.8.4.2 Manufacturer's instructions. The maximum length of the exhaust duct shall be determined by the dryer manufacturer's installation instructions. The code official shall be provided with a copy of the installation instructions for the make and model of the dryer. Where the exhaust duct is to be concealed, the installation instructions shall be provided to the code official prior to the concealment inspection. In the absence of fitting equivalent length calculations from the clothes dryer manufacturer, Table 504.8.4.1 shall be used.

504.8.4.3 Dryer exhaust duct power ventilator length. The maximum length of the exhaust duct shall be determined by the dryer exhaust duct power ventilator manufacturer's installation instructions.

504.8.5 Length identification. Where the exhaust duct equivalent length exceeds 35 feet (10 668 mm), the equivalent length of the exhaust duct shall be identified on a permanent label or tag. The label or tag shall be located within 6 feet (1829 mm) of the exhaust duct connection.

504.8.6 Exhaust duct required. Where space for a clothes dryer is provided, an exhaust duct system shall be installed. Where the clothes dryer is not installed at the time of occupancy, the exhaust duct shall be capped at the location of the future dryer.

Exception: Where a *listed* condensing clothes dryer is installed prior to occupancy of structure.

504.9 Commercial clothes dryers. The installation of dryer exhaust ducts serving commercial clothes dryers shall comply with the *appliance* manufacturer's installation instructions. Exhaust fan motors installed in exhaust systems shall be located outside of the airstream. In multiple installations, the fan shall operate continuously or be interlocked to operate when any individual unit is operating. Ducts shall have a min-

imum *clearance* of 6 inches (152 mm) to combustible materials. Clothes dryer transition ducts used to connect the *appliance* to the exhaust duct system shall be limited to single lengths not to exceed 8 feet (2438 mm) in length and shall be *listed* and *labeled* for the application. Transition ducts shall not be concealed within construction.

504.10 Common exhaust systems for clothes dryers located in multistory structures. Where a common multistory duct system is designed and installed to convey exhaust from multiple clothes dryers, the construction of the system shall be in accordance with all of the following:

1. The shaft in which the duct is installed shall be constructed and fire-resistance rated as required by the *International Building Code*.

2. Dampers shall be prohibited in the exhaust duct. Penetrations of the shaft and ductwork shall be protected in accordance with Section 607.5.5, Exception 2.

3. Rigid metal ductwork shall be installed within the shaft to convey the exhaust. The ductwork shall be constructed of sheet steel having a minimum thickness of 0.0187 inch (0.4712 mm) (No. 26 gage) and in accordance with SMACNA *Duct Construction Standards*.

4. The ductwork within the shaft shall be designed and installed without offsets.

5. The exhaust fan motor design shall be in accordance with Section 503.2.

6. The exhaust fan motor shall be located outside of the airstream.

7. The exhaust fan shall run continuously, and shall be connected to a standby power source.

8. Exhaust fan operation shall be monitored in an *approved* location and shall initiate an audible or visual signal when the fan is not in operation.

9. Makeup air shall be provided for the exhaust system.

10. A cleanout opening shall be located at the base of the shaft to provide *access* to the duct to allow for cleaning and inspection. The finished opening shall be not less than 12 inches by 12 inches (305 mm by 305 mm).

11. Screens shall not be installed at the termination.

12. The common multistory duct system shall serve only clothes dryers and shall be independent of other exhaust systems.

SECTION 505
DOMESTIC KITCHEN EXHAUST EQUIPMENT

505.1 Domestic systems. Where domestic range hoods and domestic appliances equipped with downdraft exhaust are provided, such hoods and appliances shall discharge to the outdoors through sheet metal ducts constructed of galvanized steel, stainless steel, aluminum or copper. Such ducts shall have smooth inner walls, shall be air tight, shall be equipped with a backdraft damper, and shall be independent of all other exhaust systems.

Exceptions:

1. In other than Group I-1 and I-2, where installed in accordance with the manufacturer's instructions and where mechanical or natural ventilation is otherwise provided in accordance with Chapter 4, listed and labeled ductless range hoods shall not be required to discharge to the outdoors.

2. Ducts for domestic kitchen cooking appliances equipped with downdraft exhaust systems shall be permitted to be constructed of Schedule 40 PVC pipe and fittings provided that the installation complies with all of the following:

   2.1. The duct shall be installed under a concrete slab poured on grade.

   2.2. The underfloor trench in which the duct is installed shall be completely backfilled with sand or gravel.

   2.3. The PVC duct shall extend not more than 1 inch (25 mm) above the indoor concrete floor surface.

   2.4. The PVC duct shall extend not more than 1 inch (25 mm) above grade outside of the building.

   2.5. The PVC ducts shall be solvent cemented.

505.2 Makeup air required. Exhaust hood systems capable of exhausting in excess of 400 cfm (0.19 m³/s) shall be provided with *makeup air* at a rate approximately equal to the *exhaust air* rate. Such *makeup air* systems shall be equipped with a means of closure and shall be automatically controlled to start and operate simultaneously with the exhaust system.

505.3 Common exhaust systems for domestic kitchens located in multistory structures. Where a common multistory duct system is designed and installed to convey exhaust from multiple domestic kitchen exhaust systems, the construction of the system shall be in accordance with all of the following:

1. The shaft in which the duct is installed shall be constructed and fire-resistance rated as required by the *International Building Code.*

2. Dampers shall be prohibited in the exhaust duct, except as specified in Section 505.1. Penetrations of the shaft and ductwork shall be protected in accordance with Section 607.5.5, Exception 2.

3. Rigid metal ductwork shall be installed within the shaft to convey the exhaust. The ductwork shall be constructed of sheet steel having a minimum thickness of 0.0187 inch (0.4712 mm) (No. 26 gage) and in accordance with SMACNA *Duct Construction Standards.*

4. The ductwork within the shaft shall be designed and installed without offsets.

5. The exhaust fan motor design shall be in accordance with Section 503.2.

6. The exhaust fan motor shall be located outside of the airstream.

7. The exhaust fan shall run continuously, and shall be connected to a standby power source.

8. Exhaust fan operation shall be monitored in an approved location and shall initiate an audible or visual signal when the fan is not in operation.

9. Where the exhaust rate for an individual kitchen exceeds 400 cfm (0.19 m³/s) makeup air shall be provided in accordance with Section 505.2.

10. A cleanout opening shall be located at the base of the shaft to provide access to the duct to allow for cleanout and inspection. The finished openings shall be not less than 12 inches by 12 inches (305 mm by 305 mm).

11. Screens shall not be installed at the termination.

12. The common multistory duct system shall serve only kitchen exhaust and shall be independent of other exhaust systems.

505.4 Other than Group R. In other than Group R occupancies, where domestic cooking appliances are utilized for domestic purposes, such appliances shall be provided with domestic range hoods. Hoods and exhaust systems shall be in accordance with Sections 505.1 and 505.2.

SECTION 506
COMMERCIAL KITCHEN HOOD VENTILATION SYSTEM DUCTS AND EXHAUST EQUIPMENT

506.1 General. Commercial kitchen hood ventilation ducts and exhaust *equipment* shall comply with the requirements of this section. Commercial kitchen grease ducts shall be designed for the type of cooking *appliance* and hood served.

506.2 Corrosion protection. Ducts exposed to the outside atmosphere or subject to a corrosive environment shall be protected against corrosion in an *approved* manner.

506.3 Ducts serving Type I hoods. Type I exhaust ducts shall be independent of all other exhaust systems except as provided in Section 506.3.5. Commercial kitchen duct systems serving Type I hoods shall be designed, constructed and installed in accordance with Sections 506.3.1 through 506.3.13.3.

506.3.1 Duct materials. Ducts serving Type I hoods shall be constructed of materials in accordance with Sections 506.3.1.1 and 506.3.1.2.

506.3.1.1 Grease duct materials. Grease ducts serving Type I hoods shall be constructed of steel having a minimum thickness of 0.0575 inch (1.463 mm) (No. 16 gage) or stainless steel not less than 0.0450 inch (1.14 mm) (No. 18 gage) in thickness.

Exception: Factory-built commercial kitchen grease ducts *listed* and *labeled* in accordance with UL 1978 and installed in accordance with Section 304.1.

506.3.1.2 Makeup air ducts. Makeup air ducts connecting to or within 18 inches (457 mm) of a Type I hood shall be constructed and installed in accordance with Sections 603.1, 603.3, 603.4, 603.9, 603.10 and 603.12. Duct insulation installed within 18 inches (457 mm) of a Type I hood shall be noncombustible or shall be *listed* for the application.

506.3.2 Joints, seams and penetrations of grease ducts. Joints, seams and penetrations of grease ducts shall be made with a continuous liquid-tight weld or braze made on the external surface of the duct system.

Exceptions:

1. Penetrations shall not be required to be welded or brazed where sealed by devices that are *listed* for the application.

2. Internal welding or brazing shall not be prohibited provided that the joint is formed or ground smooth and is provided with ready access for inspection.

3. Factory-built commercial kitchen grease ducts *listed* and *labeled* in accordance with UL 1978 and installed in accordance with Section 304.1.

506.3.2.1 Duct joint types. Duct joints shall be butt joints, welded flange joints with a maximum flange depth of $^1/_2$ inch (12.7 mm) or overlapping duct joints of either the telescoping or bell type. Overlapping joints shall be installed to prevent ledges and obstructions from collecting grease or interfering with gravity drainage to the intended collection point. The difference between the inside cross-sectional dimensions of overlapping sections of duct shall not exceed $^1/_4$ inch (6.4 mm). The length of overlap for overlapping duct joints shall not exceed 2 inches (51 mm).

506.3.2.2 Duct-to-hood joints. Duct-to-hood joints shall be made with continuous internal or external liquid-tight welded or brazed joints. Such joints shall be smooth, accessible for inspection, and without grease traps.

Exceptions: This section shall not apply to:

1. A vertical duct-to-hood collar connection made in the top plane of the hood in accordance with all of the following:

   1.1. The hood duct opening shall have a 1-inch-deep (25 mm), full perimeter, welded flange turned down into the hood interior at an angle of 90 degrees (1.57 rad) from the plane of the opening.

   1.2. The duct shall have a 1-inch-deep (25 mm) flange made by a 1-inch by 1-inch (25 mm by 25 mm) angle iron welded to the full perimeter of the duct not less than 1 inch (25 mm) above the bottom end of the duct.

   1.3. A gasket rated for use at not less than 1500°F (816°C) is installed between the duct flange and the top of the hood.

   1.4. The duct-to-hood joint shall be secured by stud bolts not less than $^1/_4$ inch (6.4 mm) in diameter welded to the hood with a spacing not greater than 4 inches (102 mm) on center for the full perimeter of the opening. The bolts and nuts shall be secured with lockwashers.

2. *Listed* and *labeled* duct-to-hood collar connections installed in accordance with Section 304.1.

506.3.2.3 Duct-to-exhaust fan connections. Duct-to-exhaust fan connections shall be flanged and gasketed at the base of the fan for vertical discharge fans; shall be flanged, gasketed and bolted to the inlet of the fan for side-inlet utility fans; and shall be flanged, gasketed and bolted to the inlet and outlet of the fan for in-line fans. Gasket and sealing materials shall be rated for continuous duty at a temperature of not less than 1500°F (816°C).

506.3.2.4 Vibration isolation. A vibration isolation connector for connecting a duct to a fan shall consist of noncombustible packing in a metal sleeve joint of *approved* design or shall be a coated-fabric flexible duct connector *listed* and *labeled* for the application. Vibration isolation connectors shall be installed only at the connection of a duct to a fan inlet or outlet.

506.3.2.5 Grease duct test. Prior to the use or concealment of any portion of a grease duct system, a leakage test shall be performed. Ducts shall be considered to be concealed where installed in shafts or covered by coatings or wraps that prevent the ductwork from being visually inspected on all sides. The permit holder shall be responsible to provide the necessary *equipment* and perform the grease duct leakage test. A light test shall be performed to determine that all welded and brazed joints are liquid tight.

A light test shall be performed by passing a lamp having a power rating of not less than 100 watts through the entire section of ductwork to be tested. The lamp shall be open so as to emit light equally in all directions perpendicular to the duct walls. A test shall be performed for the entire duct system, including the hood-to-duct connection. The duct work shall be permitted to be tested in sections, provided that every joint is tested. For *listed* factory-built grease ducts, this test shall be limited to duct joints assembled in the field and shall exclude factory welds.

506.3.3 Grease duct supports. Grease duct bracing and supports shall be of noncombustible material securely attached to the structure and designed to carry gravity and seismic loads within the stress limitations of the *International Building Code*. Bolts, screws, rivets and other mechanical fasteners shall not penetrate duct walls.

506.3.4 Air velocity. Grease duct systems serving a Type I hood shall be designed and installed to provide an air velocity within the duct system of not less than 500 feet per minute (2.5 m/s).

Exception: The velocity limitations shall not apply within duct transitions utilized to connect ducts to differently sized or shaped openings in hoods and fans, provided that such transitions do not exceed 3 feet (914 mm) in length and are designed to prevent the trapping of grease.

506.3.5 Separation of grease duct system. A separate grease duct system shall be provided for each Type I hood. A separate grease duct system is not required where all of the following conditions are met:

1. All interconnected hoods are located within the same story.

2. All interconnected hoods are located within the same room or in adjoining rooms.

3. Interconnecting ducts do not penetrate assemblies required to be fire-resistance rated.

4. The grease duct system does not serve solid-fuel-fired appliances.

506.3.6 Grease duct clearances. Where enclosures are not required, grease duct systems and exhaust *equipment* serving a Type I hood shall have a *clearance* to combustible construction of not less than 18 inches (457 mm), and shall have a *clearance* to noncombustible construction and gypsum wallboard attached to noncombustible structures of not less than 3 inches (76 mm).

Exceptions:

1. Factory-built commercial kitchen grease ducts *listed* and *labeled* in accordance with UL 1978.

2. *Listed* and *labeled* exhaust *equipment* installed in accordance with Section 304.1.

3. Where commercial kitchen grease ducts are continuously covered on all sides with a *listed* and *labeled* field-applied grease duct enclosure material, system, product or method of construction specifically evaluated for such purpose in accordance with ASTM E 2336, the required *clearance* shall be in accordance with the listing of such material, system, product or method.

506.3.7 Prevention of grease accumulation in grease ducts. Duct systems serving a Type I hood shall be constructed and installed so that grease cannot collect in any portion thereof, and the system shall slope not less than one-fourth unit vertical in 12 units horizontal (2-percent slope) toward the hood or toward a grease reservoir designed and installed in accordance with Section 506.3.7.1. Where horizontal ducts exceed 75 feet (22 860 mm) in length, the slope shall be not less than one unit vertical in 12 units horizontal (8.3-percent slope).

506.3.7.1 Grease duct reservoirs. Grease duct reservoirs shall:

1. Be constructed as required for the grease duct they serve.

2. Be located on the bottom of the horizontal duct or the bottommost section of the duct riser.

3. Extend across the full width of the duct and have a length of not less than 12 inches (305 mm).

4. Have a depth of not less than 1 inch (25 mm).

5. Have a bottom that slopes to a drain.

6. Be provided with a cleanout opening constructed in accordance with Section 506.3.8 and installed to provide direct access to the reservoir. The cleanout opening shall be located on a side or on top of the duct so as to permit cleaning of the reservoir.

7. Be installed in accordance with the manufacturer's instructions where manufactured devices are utilized.

506.3.8 Grease duct cleanouts and openings. Grease duct cleanouts and openings shall comply with all of the following:

1. Grease ducts shall not have openings except where required for the operation and maintenance of the system.

2. Sections of grease ducts that are inaccessible from the hood or discharge openings shall be provided with cleanout openings spaced not more than 20 feet (6096 mm) apart and not more than 10 feet (3048 mm) from changes in direction greater than 45 degrees (0.79 rad).

3. Cleanouts and openings shall be equipped with tight-fitting doors constructed of steel having a thickness not less than that required for the duct.

4. Cleanout doors shall be installed liquid tight.

5. Door assemblies including any frames and gaskets shall be approved for the application and shall not have fasteners that penetrate the duct.

6. Gasket and sealing materials shall be rated for not less than 1500°F (816°C).

7. Listed door assemblies shall be installed in accordance with the manufacturer's instructions.

506.3.8.1 Personnel entry. Where ductwork is large enough to allow entry of personnel, not less than one *approved* or *listed* opening having dimensions not less than 22 inches by 20 inches (559 mm by 508 mm) shall be provided in the horizontal sections, and in the top of vertical risers. Where such entry is provided, the duct and its supports shall be capable of supporting the additional load, and the cleanouts specified in Section 506.3.8 are not required.

506.3.8.2 Cleanouts serving in-line fans. A cleanout shall be provided for both the inlet side and outlet side of an in-line fan except where a duct does not connect to the fan. Such cleanouts shall be located within 3 feet (914 mm) of the fan duct connections.

506.3.9 Grease duct horizontal cleanouts. Cleanouts serving horizontal sections of grease ducts shall:

1. Be spaced not more than 20 feet (6096 mm) apart.

2. Be located not more than 10 feet (3048 mm) from changes in direction that are greater than 45 degrees (0.79 rad).

3. Be located on the bottom only where other locations are not available and shall be provided with internal damming of the opening such that grease will flow past the opening without pooling. Bottom cleanouts and openings shall be approved for the application and installed liquid-tight.

4. Not be closer than 1 inch (25 mm) from the edges of the duct.

5. Have opening dimensions of not less than 12 inches by 12 inches (305 mm by 305 mm). Where such dimensions preclude installation, the opening shall be not less than 12 inches (305 mm) on one side and shall be large enough to provide access for cleaning and maintenance.

6. Shall be located at grease reservoirs.

506.3.10 Underground grease duct installation. Underground grease duct installations shall comply with all of the following:

1. Underground grease ducts shall be constructed of steel having a minimum thickness of 0.0575 inch (1.463 mm) (No. 16 gage) and shall be coated to provide protection from corrosion or shall be constructed of stainless steel having a minimum thickness of 0.0450 inch (1.140 mm) (No. 18 gage).

2. The underground duct system shall be tested and approved in accordance with Section 506.3.2.5 prior to coating or placement in the ground.

3. The underground duct system shall be completely encased in concrete with a minimum thickness of 4 inches (102 mm).

4. Ducts shall slope toward grease reservoirs.

5. A grease reservoir with a cleanout to allow cleaning of the reservoir shall be provided at the base of each vertical duct riser.

6. Cleanouts shall be provided with access to permit cleaning and inspection of the duct in accordance with Section 506.3.

7. Cleanouts in horizontal ducts shall be installed on the topside of the duct.

8. Cleanout locations shall be legibly identified at the point of access from the interior space.

506.3.11 Grease duct enclosures. A commercial kitchen grease duct serving a Type I hood that penetrates a ceiling, wall, floor or any concealed space shall be enclosed from the point of penetration to the outlet terminal. In-line exhaust fans not located outdoors shall be enclosed as required for grease ducts. A duct shall penetrate exterior walls only at locations where unprotected openings are permitted by the *International Building Code*. The duct enclosure shall serve a single grease duct and shall not contain other ducts, piping or wiring systems. Duct enclosures shall be a shaft enclosure in accordance with Section 506.3.11.1, a field-applied enclosure assembly in accordance with Section 506.3.11.2 or a factory-built enclosure assembly in accordance with Section 506.3.11.3. Duct enclosures shall have a fire-resistance rating of not less than that of the assembly penetrated and not less than 1 hour. Fire dampers and smoke dampers shall not be installed in grease ducts.

Exception: A duct enclosure shall not be required for a grease duct that penetrates only a nonfire-resistance-rated roof/ceiling assembly.

506.3.11.1 Shaft enclosure. Grease ducts constructed in accordance with Section 506.3.1 shall be permitted to be enclosed in accordance with the *International Building Code* requirements for shaft construction. Such grease duct systems and exhaust *equipment* shall have a *clearance* to combustible construction of not less than 18 inches (457 mm), and shall have a *clearance* to noncombustible construction and gypsum wallboard attached to noncombustible structures of not less than 6 inches (76 mm). Duct enclosures shall be sealed around the duct at the point of penetration and vented to the outside of the building through the use of weather-protected openings.

506.3.11.2 Field-applied grease duct enclosure. Grease ducts constructed in accordance with Section 506.3.1 shall be enclosed by a *listed* and *labeled* field-applied grease duct enclosure material, systems, product, or method of construction specifically evaluated for such purpose in accordance with ASTM E 2336. The surface of the duct shall be continuously covered on all sides from the point at which the duct originates to the outlet terminal. Duct penetrations shall be protected with a through-penetration firestop system tested and *listed* in accordance with ASTM E 814 or UL 1479 and having a "F" and "T" rating equal to the fire-resistance rating of the assembly being penetrated. The grease duct enclosure and firestop system shall be installed in accordance with the listing and the manufacturer's instructions. Partial application of a field-applied grease duct enclosure shall not be installed for the sole purpose of reducing clearances to combustibles at isolated sections of grease duct. Exposed duct-wrap systems shall be protected where subject to physical damage.

506.3.11.3 Factory-built grease duct enclosure assemblies. Factory-built grease ducts incorporating integral enclosure materials shall be *listed* and *labeled* for use as grease duct enclosure assemblies specifically evaluated for such purpose in accordance with UL 2221. Duct penetrations shall be protected with a through-penetration firestop system tested and *listed* in accordance with ASTM E 814 or UL 1479 and having an "F" and "T" rating equal to the fire-resistance rating of the assembly being penetrated. The grease duct enclosure assembly and firestop system shall be installed in accordance with the listing and the manufacturer's instructions.

506.3.12 Grease duct fire-resistive access opening. Where cleanout openings are located in ducts within a fire-resistance-rated enclosure, access openings shall be provided in the enclosure at each cleanout point. Access openings shall be equipped with tight-fitting sliding or hinged doors that are equal in fire-resistive protection to that of the shaft or enclosure. An *approved* sign shall be placed on access opening panels with wording as follows: "ACCESS PANEL. DO NOT OBSTRUCT."

506.3.13 Exhaust outlets serving Type I hoods. Exhaust outlets for grease ducts serving Type I hoods shall conform to the requirements of Sections 506.3.13.1 through 506.3.13.3.

506.3.13.1 Termination above the roof. Exhaust outlets that terminate above the roof shall have the discharge opening located not less than 40 inches (1016 mm) above the roof surface.

506.3.13.2 Termination through an exterior wall. Exhaust outlets shall be permitted to terminate through exterior walls where the smoke, grease, gases, vapors and odors in the discharge from such terminations do not create a public nuisance or a fire hazard. Such terminations shall not be located where protected openings are required by the *International Building Code*. Other exterior openings shall not be located within 3 feet (914 mm) of such terminations.

506.3.13.3 Termination location. Exhaust outlets shall be located not less than 10 feet (3048 mm) horizontally from parts of the same or contiguous buildings, adjacent buildings and adjacent property lines and shall be located not less than 10 feet (3048 mm) above the adjoining grade level. Exhaust outlets shall be located not less than 10 feet (3048 mm) horizontally from or not less than 3 feet (914 mm) above air intake openings into any building.

> **Exception:** Exhaust outlets shall terminate not less than 5 feet (1524 mm) horizontally from parts of the same or contiguous building, an adjacent building, adjacent property line and air intake openings into a building where air from the exhaust outlet discharges away from such locations.

506.4 Ducts serving Type II hoods. Commercial kitchen exhaust systems serving Type II hoods shall comply with Sections 506.4.1 and 506.4.2.

506.4.1 Ducts. Ducts and plenums serving Type II hoods shall be constructed of rigid metallic materials. Duct construction, installation, bracing and supports shall comply with Chapter 6. Ducts subject to positive pressure and ducts conveying moisture-laden or waste-heat-laden air shall be constructed, joined and sealed in an *approved* manner.

506.4.2 Type II terminations. Exhaust outlets serving Type II hoods shall terminate in accordance with the hood manufacturer's installation instructions and shall comply with all of the following:

1. Exhaust outlets shall terminate not less than 3 feet (914 mm) in any direction from openings into the building.

2. Outlets shall terminate not less than 10 feet (3048 mm) from property lines or buildings on the same lot.

3. Outlets shall terminate not less than 10 feet (3048 mm) above grade.

4. Outlets that terminate above a roof shall terminate not less than 30 inches (762 mm) above the roof surface.

5. Outlets shall terminate not less than 30 inches (762 mm) from exterior vertical walls

6. Outlets shall be protected against local weather conditions.

7. Outlets shall not be directed onto walkways.

8. Outlets shall meet the provisions for exterior wall opening protectives in accordance with the *International Building Code*.

506.5 Exhaust equipment. Exhaust *equipment*, including fans and grease reservoirs, shall comply with Sections 506.5.1 through 506.5.5 and shall be of an *approved* design or shall be *listed* for the application.

506.5.1 Exhaust fans. Exhaust fan housings serving a Type I hood shall be constructed as required for grease ducts in accordance with Section 506.3.1.1.

> **Exception:** Fans *listed* and *labeled* in accordance with UL 762.

506.5.1.1 Fan motor. Exhaust fan motors shall be located outside of the exhaust airstream.

506.5.1.2 In-line fan location. Where enclosed duct systems are connected to in-line fans not located outdoors, the fan shall be located in a room or space having the same fire-resistance rating as the duct enclosure. Access shall be provided for servicing and cleaning of fan components. Such rooms or spaces shall be ventilated in accordance with the fan manufacturer's installation instructions.

506.5.2 Exhaust fan discharge. Exhaust fans shall be positioned so that the discharge will not impinge on the roof, other *equipment* or appliances or parts of the structure. A vertical discharge fan shall be manufactured with an *approved* drain outlet at the lowest point of the housing to permit drainage of grease to an *approved* grease reservoir.

506.5.3 Exhaust fan mounting. Up-blast fans serving Type I hoods and installed in a vertical or horizontal position shall be hinged, supplied with a flexible weatherproof electrical cable to permit inspection and cleaning and shall be equipped with a means of restraint to limit the swing of the fan on its hinge. The ductwork shall extend not less than 18 inches (457 mm) above the roof surface.

506.5.4 Clearances. Exhaust *equipment* serving a Type I hood shall have a *clearance* to combustible construction of not less than 18 inches (457 mm).

> **Exception:** Factory-built exhaust *equipment* installed in accordance with Section 304.1 and *listed* for a lesser *clearance*.

506.5.5 Termination location. The outlet of exhaust *equipment* serving Type I hoods shall be in accordance with Section 506.3.13.

> **Exception:** The minimum horizontal distance between vertical discharge fans and parapet-type building structures shall be 2 feet (610 mm) provided that such structures are not higher than the top of the fan discharge opening.

SECTION 507
COMMERCIAL KITCHEN HOODS

507.1 General. Commercial kitchen exhaust hoods shall comply with the requirements of this section. Hoods shall be Type I or II and shall be designed to capture and confine cooking vapors and residues. A Type I or Type II hood shall be installed at or above all *commercial cooking appliances* in accordance with Sections 507.2 and 507.3. Where any cooking *appliance* under a single hood requires a Type I hood, a Type I hood shall be installed. Where a Type II hood is required, a Type I or Type II hood shall be installed. Where a Type I hood is installed, the installation of the entire system, including the hood, ducts, exhaust equipment and makeup air system shall comply with the requirements of Sections 506, 507, 508 and 509.

Exceptions:

1. Factory-built commercial exhaust hoods that are listed and labeled in accordance with UL 710, and installed in accordance with Section 304.1, shall not be required to comply with Sections 507.1.5, 507.2.3, 507.2.5, 507.2.8, 507.3.1, 507.3.3, 507.4 and 507.5.

2. Factory-built commercial cooking recirculating systems that are listed and labeled in accordance with UL 710B, and installed in accordance with Section 304.1, shall not be required to comply with Sections 507.1.5, 507.2.3, 507.2.5, 507.2.8, 507.3.1, 507.3.3, 507.4 and 507.5. Spaces in which such systems are located shall be considered to be kitchens and shall be ventilated in accordance with Table 403.3.1.1. For the purpose of determining the floor area required to be ventilated, each individual *appliance* shall be considered as occupying not less than 100 square feet (9.3 m^2).

3. Where cooking appliances are equipped with integral down-draft exhaust systems and such appliances and exhaust systems are listed and labeled for the application in accordance with NFPA 96, a hood shall not be required at or above them.

507.1.1 Operation. Commercial kitchen exhaust hood systems shall operate during the cooking operation. The hood exhaust rate shall comply with the listing of the hood or shall comply with Section 507.5. The exhaust fan serving a Type I hood shall have automatic controls that will activate the fan when any appliance that requires such Type I Hood is turned on, or a means of interlock shall be provided that will prevent operation of such appliances when the exhaust fan is not turned on. Where one or more temperature or radiant energy sensors are used to activate a Type I hood exhaust fan, the fan shall activate not more than 15 minutes after the first appliance served by that hood has been turned on. A method of interlock between an exhaust hood system and appliances equipped with standing pilot burners shall not cause the pilot burners to be extinguished. A method of interlock between an exhaust hood system and cooking appliances shall not involve or depend upon any component of a fire-extinguishing system.

The net exhaust volumes for hoods shall be permitted to be reduced during part-load cooking conditions, where engineered or *listed* multispeed or variable speed controls automatically operate the exhaust system to maintain capture and removal of cooking effluents as required by this section. Reduced volumes shall not be below that required to maintain capture and removal of effluents from the idle cooking appliances that are operating in a standby mode.

507.1.1.1 Multiple hoods utilizing a single exhaust system. Where heat or radiant energy sensors are utilized in hood systems consisting of multiple hoods served by a single exhaust system, such sensors shall be provided in each hood. Sensors shall be capable of being accessed from the hood outlet or from a cleanout location.

507.1.2 Domestic cooking appliances used for commercial purposes. Domestic cooking appliances utilized for commercial purposes shall be provided with Type I or Type II hoods as required for the type of appliances and processes in accordance with Sections 507.2 and 507.3. Domestic cooking appliances utilized for domestic purposes shall comply with Section 505.

507.1.3 Fuel-burning appliances. Where vented fuel-burning appliances are located in the same room or space as the hood, provisions shall be made to prevent the hood system from interfering with normal operation of the *appliance* vents.

507.1.4 Cleaning. A hood shall be designed to provide for thorough cleaning of the entire hood.

507.1.5 Exhaust outlets. Exhaust outlets located within the hood shall be located so as to optimize the capture of particulate matter. Each outlet shall serve not more than a 12-foot (3658 mm) section of hood.

507.2 Type I hoods. Type I hoods shall be installed where cooking *appliances* produce grease or smoke as a result of the cooking process. Type I hoods shall be installed over *medium-duty, heavy-duty* and *extra-heavy-duty cooking appliances.*

Exception: A Type I hood shall not be required for an electric cooking appliance where an approved testing agency provides documentation that the appliance effluent contains 5 mg/m^3 or less of grease when tested at an exhaust flow rate of 500 cfm (0.236 m^3/s) in accordance with UL 710B.

507.2.1 Type I exhaust flow rate label. Type I hoods shall bear a label indicating the minimum exhaust flow rate in cfm per linear foot (1.55 L/s per linear meter) of hood that provides for capture and containment of the exhaust effluent for the cooking appliances served by the hood, based on the cooking appliance duty classifications defined in this code.

507.2.2 Type I extra-heavy-duty. Type I hoods for use over *extra-heavy-duty cooking appliances* shall not cover *heavy-, medium-* or *light-duty appliances.* Such hoods shall discharge to an exhaust system that is independent of other exhaust systems.

507.2.3 Type I materials. Type I hoods shall be constructed of steel having a minimum thickness of 0.0466 inch (1.181 mm) (No. 18 gage) or stainless steel not less than 0.0335 inch [0.8525 mm (No. 20 MSG)] in thickness.

507.2.4 Type I supports. Type I hoods shall be secured in place by noncombustible supports. Type I hood supports shall be adequate for the applied load of the hood, the unsupported ductwork, the effluent loading and the possible weight of personnel working in or on the hood.

507.2.5 Type I hoods. External hood joints, seams and penetrations for Type I hoods shall be made with a continuous external liquid-tight weld or braze to the lowest outermost perimeter of the hood. Internal hood joints, seams, penetrations, filter support frames and other appendages attached inside the hood shall not be required to be welded or brazed but shall be otherwise sealed to be grease tight.

Exceptions:

1. Penetrations shall not be required to be welded or brazed where sealed by devices that are *listed* for the application.

2. Internal welding or brazing of seams, joints and penetrations of the hood shall not be prohibited provided that the joint is formed smooth or ground so as to not trap grease, and is readily cleanable.

507.2.6 Clearances for Type I hood. A Type I hood shall be installed with a *clearance* to combustibles of not less than 18 inches (457 mm).

Exception: *Clearance* shall not be required from gypsum wallboard or $^1/_2$-inch (12.7 mm) or thicker cementitious wallboard attached to noncombustible structures provided that a smooth, cleanable, nonabsorbent and noncombustible material is installed between the hood and the gypsum or cementitious wallboard over an area extending not less than 18 inches (457 mm) in all directions from the hood.

507.2.7 Type I hoods penetrating a ceiling. Type I hoods or portions thereof penetrating a ceiling, wall or furred space shall comply with Section 506.3.11. Field-applied grease duct enclosure systems, as addressed in Section 506.3.11.2, shall not be utilized to satisfy the requirements of this section.

507.2.8 Type I grease filters. Type I hoods shall be equipped with grease filters listed and labeled in accordance with UL 1046. Grease filters shall be provided with access for cleaning or replacement. The lowest edge of a grease filter located above the cooking surface shall be not less than the height specified in Table 507.2.8.

TABLE 507.2.8
MINIMUM DISTANCE BETWEEN THE
LOWEST EDGE OF A GREASE FILTER AND
THE COOKING SURFACE OR THE HEATING SURFACE

TYPE OF COOKING APPLIANCES	HEIGHT ABOVE COOKING SURFACE (feet)
Without exposed flame	0.5
Exposed flame and burners	2
Exposed charcoal and charbroil type	3.5

For SI: 1 foot = 304.8 mm.

507.2.8.1 Criteria. Filters shall be of such size, type and arrangement as will permit the required quantity of air to pass through such units at rates not exceeding those for which the filter or unit was designed or *approved.* Filter units shall be installed in frames or holders so as to be readily removable without the use of separate tools, unless designed and installed to be cleaned in place and the system is equipped for such cleaning in place. Where filters are designed and required to be cleaned, removable filter units shall be of a size that will allow them to be cleaned in a dishwashing machine or pot sink. Filter units shall be arranged in place or provided with drip-intercepting devices to prevent grease or other condensate from dripping into food or on food preparation surfaces.

507.2.8.2 Mounting position of grease filters. Filters shall be installed at an angle of not less than 45 degrees (0.79 rad) from the horizontal and shall be equipped with a drip tray beneath the lower edge of the filters.

507.2.9 Grease gutters for Type I hood. Grease gutters shall drain to an *approved* collection receptacle that is fabricated, designed and installed to allow access for cleaning.

507.3 Type II hoods. Type II hoods shall be installed above dishwashers and appliances that produce heat or moisture and do not produce grease or smoke as a result of the cooking process, except where the heat and moisture loads from such appliances are incorporated into the HVAC system design or into the design of a separate removal system. Type II hoods shall be installed above all appliances that produce products of combustion and do not produce grease or smoke as a result of the cooking process. Spaces containing cooking appliances that do not require Type II hoods shall be provided with exhaust at a rate of 0.70 cfm per square foot (0.00033 m^3/s).

For the purpose of determining the floor area required to be exhausted, each individual appliance that is not required to be installed under a Type II hood shall be considered as occupying not less than 100 square feet (9.3 m²). Such additional square footage shall be provided with exhaust at a rate of 0.70 cfm per square foot [.00356 m³/(s × m²)].

507.3.1 Type II hood materials. Type II hoods shall be constructed of steel having a minimum thickness of 0.0296 inch (0.7534 mm) (No. 22 gage) or stainless steel not less than 0.0220 inch (0.5550 mm) (No. 24 gage) in thickness, copper sheets weighing not less than 24 ounces per square foot (7.3 kg/m²) or of other *approved* material and gage.

507.3.2 Type II supports. Type II hood supports shall be adequate for the applied load of the hood, the unsupported ductwork, the effluent loading and the possible weight of personnel working in or on the hood.

507.3.3 Type II hoods joint, seams and penetrations. Joints, seams and penetrations for Type II hoods shall be constructed as set forth in Chapter 6, shall be sealed on the interior of the hood and shall provide a smooth surface that is readily cleanable and water tight.

507.4 Hood size and location. Hoods shall comply with the overhang, setback and height requirements in accordance with Sections 507.4.1 and 507.4.2, based on the type of hood.

507.4.1 Canopy size and location. The inside lower edge of canopy-type Type I and II commercial hoods shall overhang or extend a horizontal distance of not less than 6 inches (152 mm) beyond the edge of the top horizontal surface of the *appliance* on all open sides. The vertical distance between the front lower lip of the hood and such surface shall not exceed 4 feet (1219 mm).

> **Exception:** The hood shall be permitted to be flush with the outer edge of the cooking surface where the hood is closed to the *appliance* side by a noncombustible wall or panel.

507.4.2 Noncanopy size and location. Noncanopy-type hoods shall be located not greater than 3 feet (914 mm) above the cooking surface. The edge of the hood shall be set back not greater than 1 foot (305 mm) from the edge of the cooking surface.

507.5 Capacity of hoods. Commercial food service hoods shall exhaust a minimum net quantity of air determined in accordance with this section and Sections 507.5.1 through 507.5.5. The net quantity of *exhaust air* shall be calculated by subtracting any airflow supplied directly to a hood cavity from the total exhaust flow rate of a hood. Where any combination of *heavy-duty*, *medium-duty* and *light-duty cooking appliances* are utilized under a single hood, the exhaust rate required by this section for the heaviest duty *appliance* covered by the hood shall be used for the entire hood.

507.5.1 Extra-heavy-duty cooking appliances. The minimum net airflow for hoods, as determined by Section 507.1, used for *extra-heavy-duty cooking appliances* shall be determined as follows:

Type of Hood	CFM per linear foot of hood
Backshelf/pass-over	Not allowed
Double island canopy (per side)	550
Eyebrow	Not allowed
Single island canopy	700
Wall-mounted canopy	550

For SI: 1 cfm per linear foot = 1.55 L/s per linear meter.

507.5.2 Heavy-duty cooking appliances. The minimum net airflow for hoods, as determined by Section 507.1, used for *heavy-duty cooking appliances* shall be determined as follows:

Type of Hood	CFM per linear foot of hood
Backshelf/pass-over	400
Double island canopy (per side)	400
Eyebrow	Not allowed
Single island canopy	600
Wall-mounted canopy	400

For SI: 1 cfm per linear foot = 1.55 L/s per linear meter.

507.5.3 Medium-duty cooking appliances. The minimum net airflow for hoods, as determined by Section 507.1, used for *medium-duty cooking appliances* shall be determined as follows:

Type of Hood	CFM per linear foot of hood
Backshelf/pass-over	300
Double island canopy (per side)	300
Eyebrow	250
Single island canopy	500
Wall-mounted canopy	300

For SI: 1 cfm per linear foot = 1.55 L/s per linear meter.

507.5.4 Light-duty cooking appliances. The minimum net airflow for hoods, as determined by Section 507.1, used for *light-duty* cooking appliances and food service preparation shall be determined as follows:

Type of Hood	CFM per linear foot of hood
Backshelf/pass-over	250
Double island canopy (per side)	250
Eyebrow	250
Single island canopy	400
Wall-mounted canopy	200

For SI: 1 cfm per linear foot = 1.55 L/s per linear meter.

507.5.5 Dishwashing appliances. The minimum net airflow for Type II hoods used for dishwashing appliances shall be 100 cfm per linear foot (155 L/s per linear meter) of hood length.

> **Exception:** Dishwashing appliances and *equipment* installed in accordance with Section 507.3.

507.6 Performance test. A performance test shall be conducted upon completion and before final approval of the installation of a ventilation system serving *commercial cooking appliances*. The test shall verify the rate of exhaust airflow required by Section 507.5, makeup airflow required by Section 508 and proper operation as specified in this chapter. The permit holder shall furnish the necessary test *equipment* and devices required to perform the tests.

507.6.1 Capture and containment test. The permit holder shall verify capture and containment performance of the exhaust system. This field test shall be conducted with all appliances under the hood at operating temperatures, with all sources of outdoor air providing *makeup air* for the hood operating and with all sources of recirculated air providing conditioning for the space in which the hood is located operating. Capture and containment shall be verified visually by observing smoke or steam produced by actual or simulated cooking, such as with smoke candles, smoke puffers, and similar means.

SECTION 508
COMMERCIAL KITCHEN MAKEUP AIR

508.1 Makeup air. *Makeup air* shall be supplied during the operation of commercial kitchen exhaust systems that are provided for *commercial cooking appliances*. The amount of *makeup air* supplied to the building from all sources shall be approximately equal to the amount of *exhaust air* for all exhaust systems for the building. The *makeup air* shall not reduce the effectiveness of the exhaust system. *Makeup air* shall be provided by gravity or mechanical means or both. Mechanical *makeup air* systems shall be automatically controlled to start and operate simultaneously with the exhaust system. *Makeup air* intake opening locations shall comply with Section 401.4.

508.1.1 Makeup air temperature. The temperature differential between *makeup air* and the air in the conditioned space shall not exceed 10°F (6°C) except where the added heating and cooling loads of the *makeup air* do not exceed the capacity of the HVAC system.

508.1.2 Air balance. Design plans for a facility with a commercial kitchen ventilation system shall include a schedule or diagram indicating the design outdoor air balance. The design outdoor air balance shall indicate all exhaust and replacement air for the facility, plus the net exfiltration if applicable. The total replacement air airflow rate shall equal the total exhaust airflow rate plus the net exfiltration.

508.2 Compensating hoods. Manufacturers of compensating hoods shall provide a label indicating minimum exhaust flow and/or maximum makeup airflow that provides capture and containment of the exhaust effluent.

> **Exception:** Compensating hoods with *makeup air* supplied only from the front face discharge and side face discharge openings shall not be required to be labeled with the maximum makeup airflow.

SECTION 509
FIRE SUPPRESSION SYSTEMS

509.1 Where required. *Commercial cooking appliances* required by Section 507.2 to have a Type I hood shall be provided with an *approved* automatic fire suppression system complying with the *International Building Code* and the *International Fire Code*.

SECTION 510
HAZARDOUS EXHAUST SYSTEMS

510.1 General. This section shall govern the design and construction of duct systems for hazardous exhaust and shall determine where such systems are required. Hazardous exhaust systems are systems designed to capture and control hazardous emissions generated from product handling or processes, and convey those emissions to the outdoors. Hazardous emissions include flammable vapors, gases, fumes, mists or dusts, and volatile or airborne materials posing a health hazard, such as toxic or corrosive materials. For the purposes of this section, the health-hazard rating of materials shall be as specified in NFPA 704.

For the purposes of the provisions of Section 510, a laboratory shall be defined as a facility where the use of chemicals is related to testing, analysis, teaching, research or developmental activities. Chemicals are used or synthesized on a nonproduction basis, rather than in a manufacturing process.

510.2 Where required. A hazardous exhaust system shall be required wherever operations involving the handling or processing of hazardous materials, in the absence of such exhaust systems and under normal operating conditions, have the potential to create one of the following conditions:

1. A flammable vapor, gas, fume, mist or dust is present in concentrations exceeding 25 percent of the lower flammability limit of the substance for the expected room temperature.
2. A vapor, gas, fume, mist or dust with a health-hazard rating of 4 is present in any concentration.
3. A vapor, gas, fume, mist or dust with a health-hazard rating of 1, 2 or 3 is present in concentrations exceeding 1 percent of the median lethal concentration of the substance for acute inhalation toxicity.

> **Exception:** Laboratories, as defined in Section 510.1, except where the concentrations listed in Item 1 are exceeded or a vapor, gas, fume, mist or dust with a health-hazard rating of 1, 2, 3 or 4 is present in concentrations exceeding 1 percent of the median lethal concentration of the substance for acute inhalation toxicity.

[F] 510.2.1 Lumber yards and woodworking facilities. *Equipment* or machinery located inside buildings at lumber yards and woodworking facilities which generates or emits combustible dust shall be provided with an *approved* dust-collection and exhaust system installed in accordance with this section and the *International Fire Code*. *Equipment* and systems that are used to collect, process or convey combustible dusts shall be provided with an *approved* explosion-control system.

[F] 510.2.2 Combustible fibers. *Equipment* or machinery within a building which generates or emits combustible fibers shall be provided with an *approved* dust-collecting and exhaust system. Such systems shall comply with this code and the *International Fire Code*.

510.3 Design and operation. The design and operation of the exhaust system shall be such that flammable contaminants are diluted in noncontaminated air to maintain concentrations in the exhaust flow below 25 percent of the contaminant's lower flammability limit.

510.4 Independent system. Hazardous exhaust systems shall be independent of other types of exhaust systems.

510.5 Incompatible materials and common shafts. Incompatible materials, as defined in the *International Fire Code*, shall not be exhausted through the same hazardous exhaust system. Hazardous exhaust systems shall not share common shafts with other duct systems, except where such systems are hazardous exhaust systems originating in the same fire area.

Exception: The provisions of this section shall not apply to laboratory exhaust systems where all of the following conditions apply:

1. All of the hazardous exhaust ductwork and other laboratory exhaust within both the occupied space and the shafts are under negative pressure while in operation.

2. The hazardous exhaust ductwork manifolded together within the occupied space must originate within the same fire area.

3. Hazardous exhaust ductwork originating in different fire areas and manifolded together in a common shaft shall meet the provisions of Section 717.5.3, Exception 1, Item 1.1 of the *International Building Code*.

4. Each control branch has a flow regulating device.

5. Perchloric acid hoods and connected exhaust shall be prohibited from manifolding.

6. Radioisotope hoods are equipped with filtration, carbon beds or both where required by the *registered design professional*.

7. Biological safety cabinets are filtered.

8. Each hazardous exhaust duct system shall be served by redundant exhaust fans that comply with either of the following:

    8.1. The fans shall operate simultaneously in parallel and each fan shall be individually capable of providing the required exhaust rate.

    8.2. Each of the redundant fans is controlled so as to operate when the other fan has failed or is shut down for servicing.

510.6 Design. Systems for removal of vapors, gases and smoke shall be designed by the constant velocity or equal friction methods. Systems conveying particulate matter shall be designed employing the constant velocity method.

510.6.1 Balancing. Systems conveying explosive or radioactive materials shall be prebalanced by duct sizing. Other systems shall be balanced by duct sizing with balancing devices, such as dampers. Dampers provided to balance airflow shall be provided with securely fixed minimum-position blocking devices to prevent restricting flow below the required volume or velocity.

510.6.2 Emission control. The design of the system shall be such that the emissions are confined to the area in which they are generated by air currents, hoods or enclosures and shall be exhausted by a duct system to a safe location or treated by removing contaminants.

510.6.3 Hoods required. Hoods or enclosures shall be used where contaminants originate in a limited area of a space. The design of the hood or enclosure shall be such that air currents created by the exhaust systems will capture the contaminants and transport them directly to the exhaust duct.

510.6.4 Contaminant capture and dilution. The velocity and circulation of air in work areas shall be such that contaminants are captured by an airstream at the area where the emissions are generated and conveyed into a product-conveying duct system. Contaminated air from work areas where hazardous contaminants are generated shall be diluted below the thresholds specified in Section 510.2 with air that does not contain other hazardous contaminants.

510.6.5 Makeup air. *Makeup* air shall be provided at a rate approximately equal to the rate that air is exhausted by the hazardous exhaust system. Makeup air intakes shall be located in accordance with Section 401.4.

510.6.6 Clearances. The minimum *clearance* between hoods and combustible construction shall be the *clearance* required by the duct system.

510.6.7 Ducts. Hazardous exhaust duct systems shall extend directly to the exterior of the building and shall not extend into or through ducts and plenums.

510.7 Penetrations. Penetrations of structural elements by a hazardous exhaust system shall conform to Sections 510.7.1 through 510.7.4.

Exception: Duct penetrations within Group H-5 occupancies as allowed by the *International Building Code*.

510.7.1 Fire dampers and smoke dampers. Fire dampers and smoke dampers are prohibited in hazardous exhaust ducts.

510.7.1.1 Shaft penetrations. Hazardous exhaust ducts that penetrate fire-resistance-rated shafts shall comply with Section 714.3.1 or 714.3.1.2 of the *International Building Code*.

510.7.2 Floors. Hazardous exhaust systems that penetrate a floor/ceiling assembly shall be enclosed in a fire-resistance-rated shaft constructed in accordance with the *International Building Code*.

510.7.3 Wall assemblies. Hazardous exhaust duct systems that penetrate fire-resistance-rated wall assemblies shall be enclosed in fire-resistance-rated construction from the point of penetration to the outlet terminal, except where the interior of the duct is equipped with an approved automatic fire suppression system. Ducts shall be enclosed in accordance with the *International Building Code* requirements for shaft construction and such enclosure shall have a minimum fire-resistance rating of not less than the highest fire-resistance-rated wall assembly penetrated.

510.7.4 Fire walls. Ducts shall not penetrate a fire wall.

510.8 Suppression required. Ducts shall be protected with an *approved* automatic fire suppression system installed in accordance with the *International Building Code*.

Exceptions:

1. An approved automatic fire suppression system shall not be required in ducts conveying materials, fumes, mists and vapors that are nonflammable and noncombustible under all conditions and at any concentrations.

2. Automatic fire suppression systems shall not be required in metallic and noncombustible, nonmetallic exhaust ducts in semiconductor fabrication facilities.

3. An *approved* automatic fire suppression system shall not be required in ducts where the largest cross-sectional diameter of the duct is less than 10 inches (254 mm).

4. For laboratories, as defined in Section 510.1, automatic fire protection systems shall not be required in laboratory hoods or exhaust systems

510.9 Duct construction. Ducts used to convey hazardous exhaust shall be constructed of materials *approved* for installation in such an exhaust system and shall comply with one of the following:

1. Ducts shall be constructed of *approved* G90 galvanized sheet steel, with a minimum nominal thickness as specified in Table 510.9.

2. Ducts used in systems exhausting nonflammable corrosive fumes or vapors shall be constructed of nonmetallic materials that exhibit a flame spread index of 25 or less and a smoke-developed index of 50 or less when tested in accordance with ASTM E 84 or UL 723 and that are *listed* and *labeled* for the application.

Where the products being exhausted are detrimental to the duct material, the ducts shall be constructed of alternative materials that are compatible with the exhaust.

TABLE 510.9
MINIMUM DUCT THICKNESS

DIAMETER OF DUCT OR MAXIMUM SIDE DIMENSION	MINIMUM NOMINAL THICKNESS		
	Nonabrasive materials	Nonabrasive/abrasive materials	Abrasive materials
0-8 inches	0.028 inch (No. 24 gage)	0.034 inch (No. 22 gage)	0.040 inch (No. 20 gage)
9-18 inches	0.034 inch (No. 22 gage)	0.040 inch (No. 20 gage)	0.052 inch (No. 18 gage)
19-30 inches	0.040 inch (No. 20 gage)	0.052 inch (No. 18 gage)	0.064 inch (No. 16 gage)
Over 30 inches	0.052 inch (No. 18 gage)	0.064 inch (No. 16 gage)	0.079 inch (No. 14 gage)

For SI: 1 inch = 25.4 mm.

510.9.1 Duct joints. Ducts shall be made tight with lap joints having a minimum lap of 1 inch (25 mm). Joints used in ANSI/SMACNA Round Industrial Duct Construction Standards and ANSI/SMACNA Rectangular Industrial Duct Construction Standards are also acceptable.

510.9.2 Clearance to combustibles. Ducts shall have a *clearance* to combustibles in accordance with Table 510.9.2. Exhaust gases having temperatures in excess of 600°F (316°C) shall be exhausted to a *chimney* in accordance with Section 511.2.

TABLE 510.9.2
CLEARANCE TO COMBUSTIBLES

TYPE OF EXHAUST OR TEMPERATURE OF EXHAUST (°F)	CLEARANCE TO COMBUSTIBLES (inches)
Less than 100	1
100-600	12
Flammable vapors	6

For SI: 1 inch = 25.4 mm, °C = [(°F)- 32]/1.8.

510.9.3 Explosion relief. Systems exhausting potentially explosive mixtures shall be protected with an *approved* explosion relief system or by an *approved* explosion prevention system designed and installed in accordance with NFPA 69. An explosion relief system shall be designed to minimize the structural and mechanical damage resulting from an explosion or deflagration within the exhaust system. An explosion prevention system shall be designed to prevent an explosion or deflagration from occurring.

510.10 Supports. Ducts shall be supported at intervals not exceeding 10 feet (3048 mm). Supports shall be constructed of noncombustible material.

SECTION 511
DUST, STOCK AND REFUSE CONVEYING SYSTEMS

511.1 Dust, stock and refuse conveying systems. Dust, stock and refuse conveying systems shall comply with the provisions of Section 510 and Sections 511.1.1 through 511.2.

511.1.1 Collectors and separators. Collectors and separators involving such systems as centrifugal separators, bag filter systems and similar devices, and associated supports shall be constructed of noncombustible materials and shall be located on the exterior of the building or structure. A collector or separator shall not be located nearer than 10 feet (3048 mm) to combustible construction or to an unprotected wall or floor opening, unless the collector is provided with a metal vent pipe that extends above the highest part of any roof with a distance of 30 feet (9144 mm).

Exceptions:

1. Collectors such as "Point of Use" collectors, close extraction weld fume collectors, spray finishing booths, stationary grinding tables, sanding booths, and integrated or machine-mounted collectors shall be permitted to be installed indoors provided the installation is in accordance with the *International Fire Code* and NFPA 70.

2. Collectors in independent exhaust systems handling combustible dusts shall be permitted to be installed indoors provided that such collectors are installed in compliance with the *International Fire Code* and NFPA 70.

511.1.2 Discharge pipe. Discharge piping shall conform to the requirements for ducts, including clearances required for high-heat appliances, as contained in this code. A delivery pipe from a cyclone collector shall not convey refuse directly into the firebox of a boiler, furnace, dutch oven, refuse burner, incinerator or other *appliance*.

511.1.3 Conveying systems exhaust discharge. An exhaust system shall discharge to the outside of the building either directly by flue or indirectly through the bin or vault into which the system discharges except where the contaminants have been removed. Exhaust system discharge shall be permitted to be recirculated provided that the solid particulate has been removed at a minimum efficiency of 99.9 percent at 10 microns (10.01 mm), vapor concentrations are less than 25 percent of the LFL, and *approved equipment* is used to monitor the vapor concentration.

511.1.4 Spark protection. The outlet of an open-air exhaust terminal shall be protected with an *approved* metal or other noncombustible screen to prevent the entry of sparks.

511.1.5 Explosion relief vents. A safety or explosion relief vent shall be provided on all systems that convey combustible refuse or stock of an explosive nature, in accordance with the requirements of the *International Building Code*.

511.1.5.1 Screens. Where a screen is installed in a safety relief vent, the screen shall be attached so as to permit ready release under the explosion pressure.

511.1.5.2 Hoods. The relief vent shall be provided with an *approved* noncombustible cowl or hood, or with a counterbalanced relief valve or cover arranged to prevent the escape of hazardous materials, gases or liquids.

511.2 Exhaust outlets. Outlets for exhaust that exceed 600°F (315°C) shall be designed as a *chimney* in accordance with Table 511.2.

SECTION 512
SUBSLAB SOIL EXHAUST SYSTEMS

512.1 General. Where a subslab soil exhaust system is provided, the duct shall conform to the requirements of this section.

512.2 Materials. Subslab soil exhaust system duct material shall be air duct material *listed* and *labeled* to the requirements of UL 181 for Class 0 air ducts, or any of the following

TABLE 511.2
CONSTRUCTION, CLEARANCE AND TERMINATION REQUIREMENTS FOR SINGLE-WALL METAL CHIMNEYS

CHIMNEYS SERVING	MINIMUM THICKNESS		TERMINATION				CLEARANCE			
	Walls (inch)	Lining	Above roof opening (feet)	Above any part of building within (feet)			Combustible construction (inches)		Noncombustible construction	
				10	25	50	Interior inst.	Exterior inst.	Interior inst.	Exterior inst.
High-heat appliances (Over 2,000°F)[a]	0.127 (No. 10 MSG)	4¹/₂″ laid on 4¹/₂″ bed	20	—	—	20	See Note c			
Low-heat appliances (1,000°F normal operation)	0.127 (No. 10 MSG)	none	3	2	—	—	18	6	Up to 18″ diameter, 2″ Over 18″ diameter, 4″	
Medium-heat appliances (2,000°F maximum)[b]	0.127 (No. 10 MSG)	Up to 18″ dia.—2¹/₂″ Over 18″—4¹/₂″ On 4¹/₂″ bed	10	—	10	—	36	24		

For SI: 1 inch = 25.4 mm, 1 foot = 304.8 mm, °C = [(°F)-32]/1.8.

a. Lining shall extend from bottom to top of outlet.

b. Lining shall extend from 24 inches below connector to 24 feet above.

c. Clearance shall be as specified by the design engineer and shall have sufficient clearance from buildings and structures to avoid overheating combustible materials (maximum 160°F).

piping materials that comply with the *International Plumbing Code* as building sanitary drainage and vent pipe: cast iron; galvanized steel; brass or copper pipe; copper tube of a weight not less than that of copper drainage tube, Type DWV; and plastic piping.

512.3 Grade. Exhaust system ducts shall not be trapped and shall have a minimum slope of one-eighth unit vertical in 12 units horizontal (1-percent slope).

512.4 Termination. Subslab soil exhaust system ducts shall extend through the roof and terminate not less than 6 inches (152 mm) above the roof and not less than 10 feet (3048 mm) from any operable openings or air intake.

512.5 Identification. Subslab soil exhaust ducts shall be permanently identified within each floor level by means of a tag, stencil or other *approved* marking.

SECTION 513
SMOKE CONTROL SYSTEMS

[F] 513.1 Scope and purpose. This section applies to mechanical and passive smoke control systems that are required by the *International Building Code* or the *International Fire Code*. The purpose of this section is to establish minimum requirements for the design, installation and acceptance testing of smoke control systems that are intended to provide a tenable environment for the evacuation or relocation of occupants. These provisions are not intended for the preservation of contents, the timely restoration of operations, or for assistance in fire suppression or overhaul activities. Smoke control systems regulated by this section serve a different purpose than the smoke- and heat-venting provisions found in Section 910 of the *International Building Code* or the *International Fire Code*.

[F] 513.2 General design requirements. Buildings, structures, or parts thereof required by the *International Building Code* or the *International Fire Code* to have a smoke control system or systems shall have such systems designed in accordance with the applicable requirements of Section 909 of the *International Building Code* and the generally accepted and well-established principles of engineering relevant to the design. The *construction documents* shall include sufficient information and detail to describe adequately the elements of the design necessary for the proper implementation of the smoke control systems. These documents shall be accompanied with sufficient information and analysis to demonstrate compliance with these provisions.

[F] 513.3 Special inspection and test requirements. In addition to the ordinary inspection and test requirements that buildings, structures and parts thereof are required to undergo, smoke control systems subject to the provisions of Section 909 of the *International Building Code* shall undergo special inspections and tests sufficient to verify the proper commissioning of the smoke control design in its final installed condition. The design submission accompanying the *construction documents* shall clearly detail procedures and methods to be used and the items subject to such inspections and tests. Such commissioning shall be in accordance with

generally accepted engineering practice and, where possible, based on published standards for the particular testing involved. The special inspections and tests required by this section shall be conducted under the same terms as found in Section 1704 of the *International Building Code*.

[F] 513.4 Analysis. A rational analysis supporting the types of smoke control systems to be employed, their methods of operation, the systems supporting them and the methods of construction to be utilized shall accompany the submitted *construction documents* and shall include, but not be limited to, the items indicated in Sections 513.4.1 through 513.4.7.

[F] 513.4.1 Stack effect. The system shall be designed such that the maximum probable normal or reverse stack effects will not adversely interfere with the system's capabilities. In determining the maximum probable stack effects, altitude, elevation, weather history and interior temperatures shall be used.

[F] 513.4.2 Temperature effect of fire. Buoyancy and expansion caused by the design fire in accordance with Section 513.9 shall be analyzed. The system shall be designed such that these effects do not adversely interfere with its capabilities.

[F] 513.4.3 Wind effect. The design shall consider the adverse effects of wind. Such consideration shall be consistent with the wind-loading provisions of the *International Building Code*.

[F] 513.4.4 HVAC systems. The design shall consider the effects of the heating, ventilating and air-conditioning (HVAC) systems on both smoke and fire transport. The analysis shall include all permutations of systems' status. The design shall consider the effects of fire on the HVAC systems.

[F] 513.4.5 Climate. The design shall consider the effects of low temperatures on systems, property and occupants. Air inlets and exhausts shall be located so as to prevent snow or ice blockage.

[F] 513.4.6 Duration of operation. All portions of active or engineered smoke control systems shall be capable of continued operation after detection of the fire event for a period of not less than either 20 minutes or 1.5 times the calculated egress time, whichever is greater.

513.4.7 Smoke control system interaction. The design shall consider the interaction effects of the operation of multiple smoke control systems for all design scenarios.

[F] 513.5 Smoke barrier construction. Smoke barriers required for passive smoke control and a smoke control system using the pressurization method shall comply with Section 709 of the *International Building Code*. The maximum allowable leakage area shall be the aggregate area calculated using the following leakage area ratios:

1. Walls: $A/A_w = 0.00100$

2. Interior exit stairways and ramps and exit passageways: $A/A_w = 0.00035$

3. Enclosed exit access stairways and ramps and all other shafts: $A/A_w = 0.00150$

4. Floors and roofs: $A/A_F = 0.00050$

where:

A = Total leakage area, square feet (m²).

A_F = Unit floor or roof area of barrier, square feet (m²).

A_w = Unit wall area of barrier, square feet (m²).

The leakage area ratios shown do not include openings created by gaps around doors and operable windows. The total leakage area of the smoke barrier shall be determined in accordance with Section 513.5.1 and tested in accordance with Section 513.5.2.

[F] 513.5.1 Total leakage area. Total leakage area of the barrier is the product of the smoke barrier gross area times the allowable leakage area ratio, plus the area of other openings such as gaps around doors and operable windows.

[F] 513.5.2 Testing of leakage area. Compliance with the maximum total leakage area shall be determined by achieving the minimum air pressure difference across the barrier with the system in the smoke control mode for mechanical smoke control systems utilizing the pressurization method. Compliance with the maximum total leakage area of passive smoke control systems shall be verified through methods such as door fan testing or other methods, as *approved* by the fire code official.

[F] 513.5.3 Opening protection. Openings in smoke barriers shall be protected by automatic-closing devices actuated by the required controls for the mechanical smoke control system. Door openings shall be protected by door assemblies complying with the requirements of the *International Building Code* for doors in smoke barriers.

Exceptions:

1. Passive smoke control systems with automatic-closing devices actuated by spot-type smoke detectors *listed* for releasing service installed in accordance with the *International Building Code*.

2. Fixed openings between smoke zones which are protected utilizing the airflow method.

3. In Group I-1 Condition 2, Group I-2 and ambulatory care facilities, where a pair of opposite-swinging doors are installed across a corridor in accordance with Section 513.5.3.1, the doors shall not be required to be protected in accordance with Section 716 of the *International Building Code*. The doors shall be close-fitting within operational tolerances and shall not have a center mullion or undercuts in excess of $^3/_4$ inch (19.1 mm), louvers or grilles. The doors shall have head and jamb stops and astragals or rabbets at meeting edges and, where permitted by the door manufacturer's listing, positive-latching devices are not required.

4. In Group I-2 and ambulatory care facilities, where such doors are special-purpose horizontal sliding, accordion or folding door assemblies installed in accordance with Section 1010.1.4.3 of the *International Building Code* and are automatic closing by smoke detection in accordance with Section 716.5.9.3 of the *International Building Code*.

5. Group I-3.

6. Openings between smoke zones with clear ceiling heights of 14 feet (4267 mm) or greater and bank down capacity of greater than 20 minutes as determined by the design fire size.

[F] 513.5.3.1 Group I-1 Condition 2; Group I-2 and ambulatory care facilities. In Group I-1 Condition 2; Group I-2 and *ambulatory care facilities*, where doors are installed across a *corridor*, the doors shall be automatic closing by smoke detection in accordance with Section 716.5.9.3 of the *International Building Code* and shall have a vision panel with fire-protection-rated glazing materials in fire-protection-rated frames, the area of which shall not exceed that tested.

[F] 513.5.3.2 Ducts and air transfer openings. Ducts and air transfer openings are required to be protected with a minimum Class II, 250°F (121°C) smoke damper complying with the *International Building Code*.

[F] 513.6 Pressurization method. The primary mechanical means of controlling smoke shall be by pressure differences across smoke barriers. Maintenance of a tenable environment is not required in the smoke control zone of fire origin.

[F] 513.6.1 Minimum pressure difference. The minimum pressure difference across a smoke barrier shall be 0.05-inch water gage (12.4 Pa) in fully sprinklered buildings.

In buildings permitted to be other than fully sprinklered, the smoke control system shall be designed to achieve pressure differences not less than two times the maximum calculated pressure difference produced by the design fire.

[F] 513.6.2 Maximum pressure difference. The maximum air pressure difference across a smoke barrier shall be determined by required door-opening or closing forces. The actual force required to open exit doors when the system is in the smoke control mode shall be in accordance with the *International Building Code*. Opening and closing forces for other doors shall be determined by standard engineering methods for the resolution of forces and reactions. The calculated force to set a side-hinged, swinging door in motion shall be determined by:

$$F = F_{dc} + K(WA\Delta P)/2(W-d) \qquad \textbf{(Equation 5-1)}$$

where:

A = Door area, square feet (m²).

d = Distance from door handle to latch edge of door, feet (m).

F = Total door opening force, pounds (N).

F_{dc} = Force required to overcome closing device, pounds (*N*).

K = Coefficient 5.2 (1.0).

W = Door width, feet (m).

ΔP = Design pressure difference, inches (Pa) water gage.

513.6.3 Pressurized stairways and elevator hoistways. Where stairways or elevator hoistways are pressurized, such pressurization systems shall comply with Section 513 as smoke control systems, in addition to the requirements of Sections 909.20 of the *International Building Code* and 909.21 of the *International Fire Code*.

[F] 513.7 Airflow design method. Where *approved* by the code official, smoke migration through openings fixed in a permanently open position, which are located between smoke control zones by the use of the airflow method, shall be permitted. The design airflows shall be in accordance with this section. Airflow shall be directed to limit smoke migration from the fire zone. The geometry of openings shall be considered to prevent flow reversal from turbulent effects. Smoke control systems using the airflow method shall be designed in accordance with NFPA 92.

[F] 513.7.1 Prohibited conditions. This method shall not be employed where either the quantity of air or the velocity of the airflow will adversely affect other portions of the smoke control system, unduly intensify the fire, disrupt plume dynamics or interfere with exiting. Airflow toward the fire shall not exceed 200 feet per minute (1.02 m/s). Where the calculated airflow exceeds this limit, the airflow method shall not be used.

[F] 513.8 Exhaust method. Where *approved* by the building official, mechanical smoke control for large enclosed volumes, such as in atriums or malls, shall be permitted to utilize the exhaust method. Smoke control systems using the exhaust method shall be designed in accordance with NFPA 92.

[F] 513.8.1 Exhaust rate. The height of the lowest horizontal surface of the accumulating smoke layer shall be maintained not less than 6 feet (1829 mm) above any walking surface which forms a portion of a required egress system within the smoke zone.

[F] 513.9 Design fire. The design fire shall be based on a rational analysis performed by the *registered design professional* and *approved* by the code official. The design fire shall be based on the analysis in accordance with Section 513.4 and this section.

[F] 513.9.1 Factors considered. The engineering analysis shall include the characteristics of the fuel, fuel load, effects included by the fire and whether the fire is likely to be steady or unsteady.

[F] 513.9.2 Design fire fuel. Determination of the design fire shall include consideration of the type of fuel, fuel spacing and configuration.

[F] 513.9.3 Heat-release assumptions. The analysis shall make use of the best available data from *approved* sources and shall not be based on excessively stringent limitations of combustible material.

[F] 513.9.4 Sprinkler effectiveness assumptions. A documented engineering analysis shall be provided for conditions that assume fire growth is halted at the time of sprinkler activation.

[F] 513.10 Equipment. *Equipment* such as, but not limited to, fans, ducts, automatic dampers and balance dampers shall be suitable for their intended use, suitable for the probable exposure temperatures that the rational analysis indicates, and as *approved* by the code official.

[F] 513.10.1 Exhaust fans. Components of exhaust fans shall be rated and certified by the manufacturer for the probable temperature rise to which the components will be exposed. This temperature rise shall be computed by:

$$T_s = (Q_c/mc) + (T_a) \qquad \text{(Equation 5-2)}$$

where:

c = Specific heat of smoke at smoke-layer temperature, Btu/lb°F (kJ/kg • K).

m = Exhaust rate, pounds per second (kg/s).

Q_c = Convective heat output of fire, Btu/s (kW).

T_a = Ambient temperature, °F (K).

T_s = Smoke temperature, °F (K).

Exception: Reduced T_s as calculated based on the assurance of adequate dilution air.

[F] 513.10.2 Ducts. Duct materials and joints shall be capable of withstanding the probable temperatures and pressures to which they are exposed as determined in accordance with Section 513.10.1. Ducts shall be constructed and supported in accordance with Chapter 6. Ducts shall be leak tested to 1.5 times the maximum design pressure in accordance with nationally accepted practices. Measured leakage shall not exceed 5 percent of design flow. Results of such testing shall be a part of the documentation procedure. Ducts shall be supported directly from fire-resistance-rated structural elements of the building by substantial, noncombustible supports.

Exception: Flexible connections, for the purpose of vibration isolation, that are constructed of *approved* fire-resistance-rated materials.

[F] 513.10.3 Equipment, inlets and outlets. *Equipment* shall be located so as to not expose uninvolved portions of the building to an additional fire hazard. Outdoor air inlets shall be located so as to minimize the potential for introducing smoke or flame into the building. Exhaust outlets shall be so located as to minimize reintroduction of smoke into the building and to limit exposure of the building or adjacent buildings to an additional fire hazard.

[F] 513.10.4 Automatic dampers. Automatic dampers, regardless of the purpose for which they are installed within the smoke control system, shall be *listed* and conform to the requirements of *approved* recognized standards.

[F] 513.10.5 Fans. In addition to other requirements, belt-driven fans shall have 1.5 times the number of belts required for the design duty with the minimum number of belts being two. Fans shall be selected for stable performance based on normal temperature and, where applicable, elevated temperature. Calculations and manufacturer's fan curves shall be part of the documentation procedures. Fans shall be supported and restrained by noncombustible devices in accordance with the structural design requirements of the *International Building Code*. Motors driving fans shall not be operating beyond their nameplate horsepower (kilowatts) as determined from measurement of actual current draw. Motors driving fans shall have a minimum service factor of 1.15.

[F] 513.11 Standby power. The smoke control system shall be supplied with standby power in accordance with Section 2702 of the *International Building Code*.

[F] 513.11.1 Equipment room. The standby power source and its transfer switches shall be in a room separate from the normal power transformers and switch gear and ventilated directly to and from the exterior. The room shall be enclosed with not less than 1-hour fire-resistance-rated fire barriers constructed in accordance with Section 707 of the *International Building Code* or horizontal assemblies constructed in accordance with Section 711 of the *International Building Code*, or both.

[F] 513.11.2 Power sources and power surges. Elements of the smoke management system relying on volatile memories or the like shall be supplied with uninterruptible power sources of sufficient duration to span 15-minute primary power interruption. Elements of the smoke management system susceptible to power surges shall be suitably protected by conditioners, suppressors or other *approved* means.

[F] 513.12 Detection and control systems. Fire detection systems providing control input or output signals to mechanical smoke control systems or elements thereof shall comply with the requirements of Section 907 of the *International Building Code*. Such systems shall be equipped with a control unit complying with UL 864 and listed as smoke control *equipment*.

[F] 513.12.1 Verification. Control systems for mechanical smoke control systems shall include provisions for verification. Verification shall include positive confirmation of actuation, testing, manual override and the presence of power downstream of all disconnects. A preprogrammed weekly test sequence shall report abnormal conditions audibly, visually and by printed report. The preprogrammed weekly test shall operate all devices, equipment and components used for smoke control.

Exception: Where verification of individual components tested through the preprogrammed weekly testing sequence will interfere with, and produce unwanted effects to, normal building operation, such individual components are permitted to be bypassed from the preprogrammed weekly testing, where *approved* by the building official and in accordance with both of the following:

1. Where the operation of components is bypassed from the preprogrammed weekly test, presence of power downstream of all disconnects shall be verified weekly by a listed control unit.

2. Testing of all components bypassed from the preprogrammed weekly test shall be in accordance with Section 909.20.6 of the *International Fire Code*.

[F] 513.12.2 Wiring. In addition to meeting the requirements of NFPA 70, all wiring, regardless of voltage, shall be fully enclosed within continuous raceways.

[F] 513.12.3 Activation. Smoke control systems shall be activated in accordance with the *International Building Code* or the *International Fire Code*.

[F] 513.12.4 Automatic control. Where complete automatic control is required or used, the automatic control sequences shall be initiated from an appropriately zoned automatic sprinkler system complying with Section 903.3.1.1 of the *International Fire Code*, from manual controls that are readily accessible to the fire department, and any smoke detectors required by engineering analysis.

[F] 513.13 Control-air tubing. Control-air tubing shall be of sufficient size to meet the required response times. Tubing shall be flushed clean and dry prior to final connections. Tubing shall be adequately supported and protected from damage. Tubing passing through concrete or masonry shall be sleeved and protected from abrasion and electrolytic action.

[F] 513.13.1 Materials. Control-air tubing shall be hard-drawn copper, Type L, ACR in accordance with ASTM B 42, ASTM B 43, ASTM B 68, ASTM B 88, ASTM B 251 and ASTM B 280. Fittings shall be wrought copper or brass, solder type in accordance with ASME B 16.18 or ASME B 16.22. Changes in direction shall be made with appropriate tool bends. Brass compression-type fittings shall be used at final connection to devices; other joints shall be brazed using a BCuP5 brazing alloy with solidus above 1,100°F (593°C) and liquids below 1,500°F (816°C). Brazing flux shall be used on copper-to-brass joints only.

Exception: Nonmetallic tubing used within control panels and at the final connection to devices provided all of the following conditions are met:

1. Tubing shall comply with the requirements of Section 602.2.1.3.

2. Tubing and connected device shall be completely enclosed within a galvanized or paint-grade steel enclosure having a minimum thickness of 0.0296 inch (0.7534 mm) (No. 22 gage). Entry to the enclosure shall be by copper tubing with a protective grommet of Neoprene or Teflon or by suitable brass compression to male barbed adapter.

3. Tubing shall be identified by appropriately documented coding.

4. Tubing shall be neatly tied and supported within the enclosure. Tubing bridging cabinets and doors or moveable devices shall be of sufficient length to avoid tension and excessive stress. Tubing shall be protected against abrasion. Tubing serving devices on doors shall be fastened along hinges.

[F] 513.13.2 Isolation from other functions. Control tubing serving other than smoke control functions shall be isolated by automatic isolation valves or shall be an independent system.

[F] 513.13.3 Testing. Control-air tubing shall be tested at three times the operating pressure for not less than 30 minutes without any noticeable loss in gauge pressure prior to final connection to devices.

[F] 513.14 Marking and identification. The detection and control systems shall be clearly marked at all junctions, accesses and terminations.

[F] 513.15 Control diagrams. Identical control diagrams shall be provided and maintained as required by the *International Fire Code*.

[F] 513.16 Fire fighter's smoke control panel. A fire fighter's smoke control panel for fire department emergency response purposes only shall be provided in accordance with the *International Fire Code*.

[F] 513.17 System response time. Smoke control system activation shall comply with the *International Fire Code*.

[F] 513.18 Acceptance testing. Devices, *equipment*, components and sequences shall be tested in accordance with the *International Fire Code*.

[F] 513.19 System acceptance. Acceptance of the smoke control system shall be in accordance with the *International Fire Code*.

SECTION 514
ENERGY RECOVERY VENTILATION SYSTEMS

514.1 General. Energy recovery ventilation systems shall be installed in accordance with this section. Where required for purposes of energy conservation, energy recovery ventilation systems shall comply with the *International Energy Conservation Code*. Ducted heat recovery ventilators shall be listed and labeled in accordance with UL 1812. Nonducted heat recovery ventilators shall be listed and labeled in accordance with UL 1815.

514.2 Prohibited applications. Energy recovery ventilation systems shall not be used in the following systems:

1. Hazardous exhaust systems covered in Section 510.

2. Dust, stock and refuse systems that convey explosive or flammable vapors, fumes or dust.

3. Smoke control systems covered in Section 513.

4. Commercial kitchen exhaust systems serving Type I or Type II hoods.

5. Clothes dryer exhaust systems covered in Section 504.

    Exception: The application of ERV equipment that recovers sensible heat only utilizing coil-type heat exchangers shall not be limited by this section.

514.3 Access. A means of access shall be provided to the heat exchanger and other components of the system as required for service, maintenance, repair or replacement.

514.4 Recirculated air. Air conveyed within energy recovery systems shall not be considered as recirculated air where the energy recovery ventilation system is constructed to limit cross-leakage between air streams to less than 10 percent of the total airflow design capacity.

CHAPTER 6

DUCT SYSTEMS

SECTION 601
GENERAL

601.1 Scope. Duct systems used for the movement of air in air-conditioning, heating, ventilating and exhaust systems shall conform to the provisions of this chapter except as otherwise specified in Chapters 5 and 7.

Exception: Ducts discharging combustible material directly into any *combustion* chamber shall conform to the requirements of NFPA 82.

[BF] 601.2 Air movement in egress elements. Corridors shall not serve as supply, return, exhaust, relief or *ventilation air* ducts.

Exceptions:

1. Use of a corridor as a source of *makeup air* for exhaust systems in rooms that open directly onto such corridors, including toilet rooms, bathrooms, dressing rooms, smoking lounges and janitor closets, shall be permitted, provided that each such corridor is directly supplied with outdoor air at a rate greater than the rate of *makeup air* taken from the corridor.

2. Where located within a *dwelling unit*, the use of corridors for conveying return air shall not be prohibited.

3. Where located within tenant spaces of 1,000 square feet (93 m^2) or less in area, use of corridors for conveying return air is permitted.

4. Incidental air movement from pressurized rooms within health care facilities, provided that the corridor is not the primary source of supply or return to the room.

[BF] 601.2.1 Corridor ceiling. Use of the space between the corridor ceiling and the floor or roof structure above as a return air *plenum* is permitted for one or more of the following conditions:

1. The corridor is not required to be of fire-resistance-rated construction.

2. The corridor is separated from the *plenum* by fire-resistance-rated construction.

3. The air-handling system serving the corridor is shut down upon activation of the air-handling unit smoke detectors required by this code.

4. The air-handling system serving the corridor is shut down upon detection of sprinkler waterflow where the building is equipped throughout with an automatic sprinkler system.

5. The space between the corridor ceiling and the floor or roof structure above the corridor is used as a com-

ponent of an *approved* engineered smoke control system.

[BF] 601.3 Exits. *Equipment* and ductwork for exit enclosure ventilation shall comply with one of the following items:

1. Such *equipment* and ductwork shall be located exterior to the building and shall be directly connected to the exit enclosure by ductwork enclosed in construction as required by the *International Building Code* for shafts.

2. Where such *equipment* and ductwork is located within the exit enclosure, the intake air shall be taken directly from the outdoors and the *exhaust air* shall be discharged directly to the outdoors, or such air shall be conveyed through ducts enclosed in construction as required by the *International Building Code* for shafts.

3. Where located within the building, such *equipment* and ductwork shall be separated from the remainder of the building, including other mechanical *equipment*, with construction as required by the *International Building Code* for shafts.

In each case, openings into fire-resistance-rated construction shall be limited to those needed for maintenance and operation and shall be protected by self-closing fire-resistance-rated devices in accordance with the *International Building Code* for enclosure wall opening protectives. Exit enclosure ventilation systems shall be independent of other building ventilation systems.

601.4 Contamination prevention. Exhaust ducts under positive pressure, chimneys and vents shall not extend into or pass through ducts or plenums.

Exceptions:

1. Exhaust systems located in ceiling return air plenums over spaces that are permitted to have 10 percent recirculation in accordance with Section 403.2.1, Item 4. The exhaust duct joints, seams and connections shall comply with Section 603.9.

2. This section shall not apply to chimneys and vents that pass through plenums where such venting systems comply with one of the following requirements:

   2.1. The venting system shall be listed for positive pressure applications and shall be sealed in accordance with the vent manufacturer's instructions.

   2.2. The venting system shall be installed such that fittings and joints between sections are not installed in the above ceiling space.

   2.3. The venting system shall be installed in a conduit or enclosure with sealed joints separating the interior of the conduit or enclosure from the ceiling space.

601.5 Return air openings. Return air openings for heating, ventilation and air-conditioning systems shall comply with all of the following:

1. Openings shall not be located less than 10 feet (3048 mm) measured in any direction from an open combustion chamber or draft hood of another appliance located in the same room or space.

2. Return air shall not be taken from a hazardous or insanitary location or a refrigeration room as defined in this code.

3. The amount of return air taken from any room or space shall be not greater than the flow rate of supply air delivered to such room or space.

4. Return and transfer openings shall be sized in accordance with the appliance or equipment manufacturer's installation instructions, ACCA Manual D or the design of the registered design professional.

5. Return air taken from one dwelling unit shall not be discharged into another dwelling unit.

6. Taking return air from a crawl space shall not be accomplished through a direct connection to the return side of a forced air furnace. Transfer openings in the crawl space enclosure shall not be prohibited.

7. Return air shall not be taken from a closet, bathroom, toilet room, kitchen, garage, boiler room, furnace room or unconditioned attic.

Exceptions:

1. Taking return air from a kitchen is not prohibited where such return air openings serve the kitchen and are located not less than 10 feet (3048 mm) from the cooking appliances.

2. Dedicated forced air systems serving only the garage shall not be prohibited from obtaining return air from the garage.

SECTION 602
PLENUMS

602.1 General. Supply, return, exhaust, relief and ventilation air plenums shall be limited to uninhabited crawl spaces, areas above a ceiling or below the floor, attic spaces and mechanical equipment rooms. Plenums shall be limited to one fire area. Air systems shall be ducted from the boundary of the fire area served directly to the air-handling equipment. Fuel-fired appliances shall not be installed within a plenum.

602.2 Construction. *Plenum* enclosure construction materials that are exposed to the airflow shall comply with the requirements of Section 703.5 of the *International Building Code* or such materials shall have a flame spread index of not more than 25 and a smoke-developed index of not more than 50 when tested in accordance with ASTM E 84 or UL 723.

The use of gypsum boards to form plenums shall be limited to systems where the air temperatures do not exceed 125°F (52°C) and the building and mechanical system design conditions are such that the gypsum board surface tempera-

ture will be maintained above the airstream dew-point temperature. Air plenums formed by gypsum boards shall not be incorporated in air-handling systems utilizing evaporative coolers.

602.2.1 Materials within plenums. Except as required by Sections 602.2.1.1 through 602.2.1.7, materials within plenums shall be noncombustible or shall be listed and labeled as having a flame spread index of not more than 25 and a smoke-developed index of not more than 50 when tested in accordance with ASTM E 84 or UL 723.

Exceptions:

1. Rigid and flexible ducts and connectors shall conform to Section 603.

2. Duct coverings, linings, tape and connectors shall conform to Sections 603 and 604.

3. This section shall not apply to materials exposed within plenums in one- and two-family dwellings.

4. This section shall not apply to smoke detectors.

5. Combustible materials fully enclosed within one of the following:

   5.1. Continuous noncombustible raceways or enclosures.

   5.2. Approved gypsum board assemblies.

   5.3. Materials listed and labeled for installation within a plenum.

6. Materials in Group H, Division 5 fabrication areas and the areas above and below the fabrication area that share a common air recirculation path with the fabrication area.

602.2.1.1 Wiring. Combustible electrical wires and cables and optical fiber cables exposed within a plenum shall be listed as having a maximum peak optical density of 0.50 or less, an average optical density of 0.15 or less, and a maximum flame spread distance of 5 feet (1524 mm) or less when tested in accordance with NFPA 262 or shall be installed in metal raceways or metal sheathed cable. Combustible optical fiber and communication raceways exposed within a plenum shall be listed as having a maximum peak optical density of 0.5 or less, an average optical density of 0.15 or less, and a maximum flame spread distance of 5 feet (1524 mm) or less when tested in accordance with ANSI/UL 2024. Only plenum-rated wires and cables shall be installed in plenum-rated raceways. Electrical wires and cables, optical fiber cables and raceways addressed in this section shall be listed and labeled and shall be installed in accordance with NFPA 70.

602.2.1.2 Fire sprinkler piping. Plastic fire sprinkler piping exposed within a *plenum* shall be used only in wet pipe systems and shall have a peak optical density not greater than 0.50, an average optical density not greater than 0.15, and a flame spread of not greater than 5 feet (1524 mm) when tested in accordance with UL 1887. Piping shall be *listed* and *labeled*.

602.2.1.3 Pneumatic tubing. Combustible pneumatic tubing exposed within a *plenum* shall have a peak optical density not greater than 0.50, an average optical density not greater than 0.15, and a flame spread of not greater than 5 feet (1524 mm) when tested in accordance with UL 1820. Combustible pneumatic tubing shall be *listed* and *labeled*.

602.2.1.4 Electrical equipment in plenums. Electrical *equipment* exposed within a *plenum* shall comply with Sections 602.2.1.4.1 and 602.2.1.4.2.

602.2.1.4.1 Equipment in metallic enclosures. Electrical *equipment* with metallic enclosures exposed within a *plenum* shall be permitted.

602.2.1.4.2 Equipment in combustible enclosures. Electrical *equipment* with combustible enclosures exposed within a *plenum* shall be *listed* and *labeled* for such use in accordance with UL 2043.

602.2.1.5 Discrete plumbing and mechanical products in plenums. Where discrete plumbing and mechanical products and appurtenances are located in a plenum and have exposed combustible material, they shall be listed and labeled for such use in accordance with UL 2043.

602.2.1.6 Foam plastic insulation. Foam plastic insulation used in plenums as interior wall or ceiling finish or as interior trim shall exhibit a flame spread index of 75 or less and a smoke-developed index of 450 or less when tested in accordance with ASTM E 84 or UL 723 and shall also comply with one or more of Sections 602.2.1.6.1, 602.2.1.6.2 and 602.2.1.6.3.

602.2.1.6.1 Separation required. The foam plastic insulation shall be separated from the plenum by a thermal barrier complying with Section 2603.4 of the *International Building Code* and shall exhibit a flame spread index of 75 or less and a smoke-developed index of 450 or less when tested in accordance with ASTM E 84 or UL 723 at the thickness and density intended for use.

602.2.1.6.2 Approval. The foam plastic insulation shall exhibit a flame spread index of 25 or less and a smoke-developed index of 50 or less when tested in accordance with ASTM E 84 or UL 723 at the thickness and density intended for use and shall meet the acceptance criteria of Section 803.1.2 of the *International Building Code* when tested in accordance with NFPA 286.

The foam plastic insulation shall be approved based on tests conducted in accordance with Section 2603.9 of the *International Building Code*.

602.2.1.6.3 Covering. The foam plastic insulation shall be covered by corrosion-resistant steel having a base metal thickness of not less than 0.0160 inch (0.4 mm) and shall exhibit a flame spread index of 75 or less and a smoke-developed index of 450 or less when tested in accordance with ASTM E 84 or UL 723 at the thickness and density intended for use.

602.2.1.7 Plastic plumbing pipe and tube. Plastic piping and tubing used in plumbing systems shall be listed and shall exhibit a flame spread index of not more than 25 and a smoke-developed index of not more than 50 when tested in accordance with ASTM E 84 or UL 723.

602.3 Stud cavity and joist space plenums. Stud wall cavities and the spaces between solid floor joists to be utilized as air plenums shall comply with the following conditions:

1. Such cavities or spaces shall not be utilized as a *plenum* for supply air.
2. Such cavities or spaces shall not be part of a required fire-resistance-rated assembly.
3. Stud wall cavities shall not convey air from more than one floor level.
4. Stud wall cavities and joist space plenums shall comply with the floor penetration protection requirements of the *International Building Code*.
5. Stud wall cavities and joist space plenums shall be isolated from adjacent concealed spaces by *approved* fireblocking as required in the *International Building Code*.
6. Stud wall cavities in the outside walls of building envelope assemblies shall not be utilized as air plenums.

[BS] 602.4 Flood hazard. For structures located in flood hazard areas, plenum spaces shall be located above the elevation required by Section 1612 of the *International Building Code* for utilities and attendant equipment or shall be designed and constructed to prevent water from entering or accumulating within the plenum spaces during floods up to such elevation. If the plenum spaces are located below the elevation required by Section 1612 of the *International Building Code* for utilities and attendant equipment, they shall be capable of resisting hydrostatic and hydrodynamic loads and stresses, including the effects of buoyancy, during the occurrence of flooding up to such elevation.

SECTION 603
DUCT CONSTRUCTION AND INSTALLATION

603.1 General. An air distribution system shall be designed and installed to supply the required distribution of air. The installation of an air distribution system shall not affect the fire protection requirements specified in the *International Building Code*. Ducts shall be constructed, braced, reinforced and installed to provide structural strength and durability.

603.2 Duct sizing. Ducts installed within a single dwelling unit shall be sized in accordance with ACCA Manual D, the appliance manufacturer's installation instructions or other approved methods. Ducts installed within all other buildings shall be sized in accordance with the ASHRAE *Handbook of Fundamentals* or other equivalent computation procedure.

603.3 Duct classification. Ducts shall be classified based on the maximum operating pressure of the duct at pressures of positive or negative 0.5, 1.0, 2.0, 3.0, 4.0, 6.0 or 10.0 inches (1 inch w.c. = 248.7 Pa) of water column. The pressure classification of ducts shall equal or exceed the design pressure of the air distribution in which the ducts are utilized.

603.4 Metallic ducts. Metallic ducts shall be constructed as specified in the SMACNA *HVAC Duct Construction Standards—Metal and Flexible.*

Exception: Ducts installed within single *dwelling units* shall have a minimum thickness as specified in Table 603.4.

603.4.1 Minimum fasteners. Round metallic ducts shall be mechanically fastened by means of not less than three sheet metal screws or rivets spaced equally around the joint.

Exception: Where a duct connection is made that is partially inaccessible, three screws or rivets shall be equally spaced on the exposed portion so as to prevent a hinge effect.

603.4.2 Duct lap. Crimp joints for round and oval metal ducts shall be lapped not less than 1 inch (25 mm) and the male end of the duct shall extend into the adjoining duct in the direction of airflow.

603.5 Nonmetallic ducts. Nonmetallic ducts shall be constructed with Class 0 or Class 1 duct material and shall comply with UL 181. Fibrous duct construction shall conform to the SMACNA *Fibrous Glass Duct Construction Standards* or NAIMA *Fibrous Glass Duct Construction Standards.* The air temperature within nonmetallic ducts shall not exceed 250ºF (121ºC).

603.5.1 Gypsum ducts. The use of gypsum boards to form air shafts (ducts) shall be limited to return air systems where the air temperatures do not exceed 125ºF (52ºC) and the gypsum board surface temperature is maintained above the airstream dew-point temperature. Air ducts formed by gypsum boards shall not be incorporated in air-handling systems utilizing evaporative coolers.

603.6 Flexible air ducts and flexible air connectors. Flexible air ducts, both metallic and nonmetallic, shall comply with Sections 603.6.1, 603.6.1.1, 603.6.3 and 603.6.4. Flexible air connectors, both metallic and nonmetallic, shall comply with Sections 603.6.2 through 603.6.4.

603.6.1 Flexible air ducts. Flexible air ducts, both metallic and nonmetallic, shall be tested in accordance with UL 181. Such ducts shall be *listed* and *labeled* as Class 0 or Class 1 flexible air ducts and shall be installed in accordance with Section 304.1.

603.6.1.1 Duct length. Flexible air ducts shall not be limited in length.

603.6.2 Flexible air connectors. Flexible air connectors, both metallic and nonmetallic, shall be tested in accordance with UL 181. Such connectors shall be *listed* and *labeled* as Class 0 or Class 1 flexible air connectors and shall be installed in accordance with Section 304.1.

603.6.2.1 Connector length. Flexible air connectors shall be limited in length to 14 feet (4267 mm).

603.6.2.2 Connector penetration limitations. Flexible air connectors shall not pass through any wall, floor or ceiling.

603.6.3 Air temperature. The design temperature of air to be conveyed in flexible air ducts and flexible air connectors shall be less than 250ºF (121ºC).

603.6.4 Flexible air duct and air connector clearance. Flexible air ducts and air connectors shall be installed with a minimum *clearance* to an *appliance* as specified in the *appliance* manufacturer's installation instructions.

TABLE 603.4
DUCT CONSTRUCTION MINIMUM SHEET METAL THICKNESS FOR SINGLE DWELLING UNITS[a]

ROUND DUCT DIAMETER (inches)	STATIC PRESSURE			
	1/2-inch water gage		1-inch water gage	
	Thickness (inches)		Thickness (inches)	
	Galvanized	Aluminum	Galvanized	Aluminum
< 12	0.013	0.018	0.013	0.018
12 to14	0.013	0.018	0.016	0.023
15 to 17	0.016	0.023	0.019	0.027
18	0.016	0.023	0.024	0.034
19 to 20	0.019	0.027	0.024	0.034

RECTANGULAR DUCT DIMENSION (inches)	STATIC PRESSURE			
	1/2-inch water gage		1-inch water gage	
	Thickness (inches)		Thickness (inches)	
	Galvanized	Aluminum	Galvanized	Aluminum
≤ 8	0.013	0.018	0.013	0.018
9 to10	0.013	0.018	0.016	0.023
11 to 12	0.016	0.023	0.019	0.027
13 to16	0.019	0.027	0.019	0.027
17 to 18	0.019	0.027	0.024	0.034
19 to 20	0.024	0.034	0.024	0.034

For SI: 1 inch = 25.4 mm, 1-inch water gage = 249 Pa.

a. Ductwork that exceeds 20 inches by dimension or exceeds a pressure of 1-inch water gage shall be constructed in accordance with SMACNA *HVAC Duct Construction Standards—Metal and Flexible.*

603.7 Rigid duct penetrations. Duct system penetrations of walls, floors, ceilings and roofs and air transfer openings in such building components shall be protected as required by Section 607. Ducts in a private garage that penetrate a wall or ceiling that separates a dwelling from a private garage shall be continuous, shall be constructed of sheet steel having a thickness of not less than 0.0187 inch (0.4712 mm) (No. 26 gage) and shall not have openings into the garage. Fire and smoke dampers are not required in such ducts passing through the wall or ceiling separating a dwelling from a private garage except where required by Chapter 7 of the *International Building Code.*

603.8 Underground ducts. Ducts shall be *approved* for underground installation. Metallic ducts not having an *approved* protective coating shall be completely encased in not less than 2 inches (51 mm) of concrete.

603.8.1 Slope. Ducts shall have a minimum slope of $^1/_8$ inch per foot (10.4 mm/m) to allow drainage to a point provided with access.

603.8.2 Sealing. Ducts shall be sealed and secured prior to pouring the concrete encasement.

603.8.3 Plastic ducts and fittings. Plastic ducts shall be constructed of PVC having a minimum pipe stiffness of 8 psi (55 kPa) at 5-percent deflection when tested in accordance with ASTM D 2412. Plastic duct fittings shall be constructed of either PVC or high-density polyethylene. Plastic duct and fittings shall be utilized in underground installations only. The maximum design temperature for systems utilizing plastic duct and fittings shall be 150°F (66°C).

603.9 Joints, seams and connections. All longitudinal and transverse joints, seams and connections in metallic and nonmetallic ducts shall be constructed as specified in SMACNA *HVAC Duct Construction Standards—Metal and Flexible* and NAIMA *Fibrous Glass Duct Construction Standards.* All joints, longitudinal and transverse seams and connections in ductwork shall be securely fastened and sealed with welds, gaskets, mastics (adhesives), mastic-plus-embedded-fabric systems, liquid sealants or tapes. Tapes and mastics used to seal fibrous glass ductwork shall be listed and labeled in accordance with UL 181A and shall be marked "181 A-P" for pressure-sensitive tape, "181 A-M" for mastic or "181 A-H" for heat-sensitive tape. Tapes and mastics used to seal metallic and flexible air ducts and flexible air connectors shall comply with UL 181B and shall be marked "181 B-FX" for pressure-sensitive tape or "181 B-M" for mastic. Duct connections to flanges of air distribution system equipment shall be sealed and mechanically fastened. Mechanical fasteners for use with flexible nonmetallic air ducts shall comply with UL 181B and shall be marked "181 B-C." Closure systems used to seal all ductwork shall be installed in accordance with the manufacturer's instructions.

Exception: For ducts having a static pressure classification of less than 2 inches of water column (500 Pa), additional closure systems shall not be required for continuously welded joints and seams and locking-type joints and seams of other than the snap-lock and button-lock types.

603.10 Supports. Ducts shall be supported in accordance with SMACNA *HVAC Duct Construction Standards—Metal and Flexible.* Flexible and other factory-made ducts shall be supported in accordance with the manufacturer's instructions.

603.11 Furnace connections. Ducts connecting to a furnace shall have a *clearance* to combustibles in accordance with the furnace manufacturer's installation instructions.

603.12 Condensation. Provisions shall be made to prevent the formation of condensation on the exterior of any duct.

[BS] 603.13 Flood hazard areas. For structures in flood hazard areas, ducts shall be located above the elevation required by Section 1612 of the *International Building Code* for utilities and attendant equipment or shall be designed and constructed to prevent water from entering or accumulating within the ducts during floods up to such elevation. If the ducts are located below the elevation required by Section 1612 of the *International Building Code* for utilities and attendant equipment, the ducts shall be capable of resisting hydrostatic and hydrodynamic loads and stresses, including the effects of buoyancy, during the occurrence of flooding up to such elevation.

603.14 Location. Ducts shall not be installed in or within 4 inches (102 mm) of the earth, except where such ducts comply with Section 603.8.

603.15 Mechanical protection. Ducts installed in locations where they are exposed to mechanical damage by vehicles or from other causes shall be protected by *approved* barriers.

603.16 Weather protection. Ducts including linings, coverings and vibration isolation connectors installed on the exterior of the building shall be protected against the elements.

603.17 Air dispersion systems. Air dispersion systems shall:

1. Be installed entirely in exposed locations.

2. Be utilized in systems under positive pressure.

3. Not pass through or penetrate fire-resistant-rated construction.

4. Be listed and labeled in compliance with UL 2518.

603.18 Registers, grilles and diffusers. Duct registers, grilles and diffusers shall be installed in accordance with the manufacturer's instructions. Volume dampers or other means of supply air adjustment shall be provided in the branch ducts or at each individual duct register, grille or diffuser. Each volume damper or other means of supply air adjustment used in balancing shall be provided with access.

603.18.1 Floor registers. Floor registers shall resist, without structural failure, a 200-pound (90.8 kg) concentrated load on a 2-inch-diameter (51 mm) disc applied to the most critical area of the exposed face.

603.18.2 Prohibited locations. Diffusers, registers and grilles shall be prohibited in the floor or its upward extension within toilet and bathing rooms required by the *International Building Code* to have smooth, hard, nonabsorbent surfaces.

Exception: *Dwelling units.*

SECTION 604
INSULATION

604.1 General. Duct insulation shall conform to the requirements of Sections 604.2 through 604.13 and the *International Energy Conservation Code*.

604.2 Surface temperature. Ducts that operate at temperatures exceeding 120°F (49°C) shall have sufficient thermal insulation to limit the exposed surface temperature to 120°F (49°C).

604.3 Coverings and linings. Coverings and linings, including adhesives where used, shall have a flame spread index not more than 25 and a smoke-developed index not more than 50, when tested in accordance with ASTM E 84 or UL 723, using the specimen preparation and mounting procedures of ASTM E 2231. Duct coverings and linings shall not flame, glow, smolder or smoke when tested in accordance with ASTM C 411 at the temperature to which they are exposed in service. The test temperature shall not fall below 250°F (121°C). Coverings and linings shall be listed and labeled.

604.4 Foam plastic insulation. Foam plastic used as duct coverings and linings shall conform to the requirements of Section 604.

604.5 Appliance insulation. *Listed* and *labeled* appliances that are internally insulated shall be considered as conforming to the requirements of Section 604.

604.6 Penetration of assemblies. Duct coverings shall not penetrate a wall or floor required to have a fire-resistance rating or required to be fireblocked.

604.7 Identification. External duct insulation, except spray polyurethane foam, and factory-insulated flexible duct shall be legibly printed or identified at intervals not greater than 36 inches (914 mm) with the name of the manufacturer, the thermal resistance *R*-value at the specified installed thickness and the flame spread and smoke-developed indexes of the composite materials. Duct insulation product *R*-values shall be based on insulation only, excluding air films, vapor retarders or other duct components, and shall be based on tested *C*-values at 75°F (24°C) mean temperature at the installed thickness, in accordance with recognized industry procedures. The installed thickness of duct insulation used to determine its *R*-value shall be determined as follows:

1. For duct board, duct liner and factory-made rigid ducts not normally subjected to compression, the nominal insulation thickness shall be used.

2. For duct wrap, the installed thickness shall be assumed to be 75 percent (25 percent compression) of nominal thickness.

3. For factory-made flexible air ducts, the installed thickness shall be determined by dividing the difference between the actual outside diameter and nominal inside diameter by two.

4. For spray polyurethane foam, the aged *R*-value per inch, measured in accordance with recognized industry standards, shall be provided to the customer in writing at the time of foam application.

604.8 Lining installation. Linings shall be interrupted at the area of operation of a fire damper and at not less than 6 inches (152 mm) upstream of and 6 inches (152 mm) downstream of electric-resistance and fuel-burning heaters in a duct system. Metal nosings or sleeves shall be installed over exposed duct liner edges that face opposite the direction of airflow.

604.9 Thermal continuity. Where a duct liner has been interrupted, a duct covering of equal thermal performance shall be installed.

604.10 Service openings. Service openings shall not be concealed by duct coverings unless the exact location of the opening is properly identified.

604.11 Vapor retarders. Where ducts used for cooling are externally insulated, the insulation shall be covered with a vapor retarder having a maximum permeance of 0.05 perm [2.87 ng/(Pa • s • m^2)] or aluminum foil having a minimum thickness of 2 mils (0.051 mm). Insulations having a permeance of 0.05 perm [2.87 ng/(Pa • s • m^2)] or less shall not be required to be covered. All joints and seams shall be sealed to maintain the continuity of the vapor retarder.

604.12 Weatherproof barriers. Insulated exterior ducts shall be protected with an *approved* weatherproof barrier.

604.13 Internal insulation. Materials used as internal insulation and exposed to the airstream in ducts shall be shown to be durable when tested in accordance with UL 181. Exposed internal insulation that is not impermeable to water shall not be used to line ducts or plenums from the exit of a cooling coil to the downstream end of the drain pan.

SECTION 605
AIR FILTERS

605.1 General. Heating and air-conditioning systems shall be provided with *approved* air filters. Filters shall be installed such that all return air, outdoor air and makeup air is filtered upstream from any heat exchanger or coil. Filters shall be installed in an *approved* convenient location. Liquid adhesive coatings used on filters shall have a flash point not lower than 325°F (163°C).

605.2 Approval. Media-type and electrostatic-type air filters shall be *listed* and *labeled*. Media-type air filters shall comply with UL 900. High efficiency particulate air filters shall comply with UL 586. Electrostatic-type air filters shall comply with UL 867. Air filters utilized within *dwelling units* shall be designed for the intended application and shall not be required to be *listed* and *labeled*.

605.3 Airflow over the filter. Ducts shall be constructed to allow an even distribution of air over the entire filter.

SECTION 606
SMOKE DETECTION SYSTEMS CONTROL

606.1 Controls required. Air distribution systems shall be equipped with smoke detectors *listed* and *labeled* for installation in air distribution systems, as required by this section. Duct smoke detectors shall comply with UL 268A. Other smoke detectors shall comply with UL 268.

606.2 Where required. Smoke detectors shall be installed where indicated in Sections 606.2.1 through 606.2.3.

> **Exception:** Smoke detectors shall not be required where air distribution systems are incapable of spreading smoke beyond the enclosing walls, floors and ceilings of the room or space in which the smoke is generated.

606.2.1 Return air systems. Smoke detectors shall be installed in return air systems with a design capacity greater than 2,000 cfm (0.9 m³/s), in the return air duct or *plenum* upstream of any filters, *exhaust air* connections, outdoor air connections, or decontamination *equipment* and appliances.

> **Exception:** Smoke detectors are not required in the return air system where all portions of the building served by the air distribution system are protected by area smoke detectors connected to a fire alarm system in accordance with the *International Fire Code*. The area smoke detection system shall comply with Section 606.4.

606.2.2 Common supply and return air systems. Where multiple air-handling systems share common supply or return air ducts or plenums with a combined design capacity greater than 2,000 cfm (0.9 m³/s), the return air system shall be provided with smoke detectors in accordance with Section 606.2.1.

> **Exception:** Individual smoke detectors shall not be required for each fan-powered terminal unit, provided that such units do not have an individual design capacity greater than 2,000 cfm (0.9 m³/s) and will be shut down by activation of one of the following:
>
> 1. Smoke detectors required by Sections 606.2.1 and 606.2.3.
>
> 2. An *approved* area smoke detector system located in the return air *plenum* serving such units.
>
> 3. An area smoke detector system as prescribed in the exception to Section 606.2.1.
>
> In all cases, the smoke detectors shall comply with Sections 606.4 and 606.4.1.

606.2.3 Return air risers. Where return air risers serve two or more stories and serve any portion of a return air system having a design capacity greater than 15,000 cfm (7.1 m³/s), smoke detectors shall be installed at each story. Such smoke detectors shall be located upstream of the connection between the return air riser and any air ducts or plenums.

[F] 606.3 Installation. Smoke detectors required by this section shall be installed in accordance with NFPA 72. The required smoke detectors shall be installed to monitor the entire airflow conveyed by the system including return air and exhaust or relief air. Access shall be provided to smoke detectors for inspection and maintenance.

[F] 606.4 Controls operation. Upon activation, the smoke detectors shall shut down all operational capabilities of the air distribution system in accordance with the listing and labeling of appliances used in the system. Air distribution systems that are part of a smoke control system shall switch to the smoke control mode upon activation of a detector.

[F] 606.4.1 Supervision. The duct smoke detectors shall be connected to a fire alarm system where a fire alarm system is required by Section 907.2 of the *International Fire Code*. The actuation of a duct smoke detector shall activate a visible and audible supervisory signal at a constantly attended location. In facilities that are required to be monitored by a supervising station, duct smoke detectors shall report only as a supervisory signal, not as a fire alarm.

> **Exceptions:**
>
> 1. The supervisory signal at a constantly attended location is not required where the duct smoke detector activates the building's alarm-indicating appliances.
>
> 2. In occupancies not required to be equipped with a fire alarm system, actuation of a smoke detector shall activate a visible and audible signal in an *approved* location. Duct smoke detector trouble conditions shall activate a visible or audible signal in an *approved* location and shall be identified as air duct detector trouble.

SECTION 607
DUCT AND TRANSFER OPENINGS

[BF] 607.1 General. The provisions of this section shall govern the protection of duct penetrations and air transfer openings in assemblies required to be protected.

[BF] 607.1.1 Ducts and air transfer openings. Ducts transitioning horizontally between shafts shall not require a shaft enclosure provided that the duct penetration into each associated shaft is protected with dampers complying with this section.

[BF] 607.1.2 Ducts that penetrate fire-resistance-rated assemblies without dampers. Ducts that penetrate fire-resistance-rated assemblies and are not required by this section to have dampers shall comply with the requirements of Sections 714.2 through 714.3.3 of the *International Building Code*. Ducts that penetrate horizontal assemblies not required to be contained within a shaft and not required by this section to have dampers shall comply with the requirements of Sections 714.4 of the *International Building Code*.

[BF] 607.1.2.1 Ducts that penetrate nonfire-resistance-rated assemblies. The space around a duct penetrating a nonfire-resistance-rated floor assembly shall comply with Section 717.6.3 of the *International Building Code*.

[BF] 607.2 Installation. Fire dampers, smoke dampers, combination fire/smoke dampers and ceiling radiation dampers located within air distribution and smoke control systems shall be installed in accordance with the requirements of this section, and the manufacturer's instructions and listing.

[BF] 607.2.1 Smoke control system. Where the installation of a fire damper will interfere with the operation of a required smoke control system in accordance with Section 909 of the *International Building Code, approved* alternative protection shall be used. Where mechanical systems including ducts and dampers used for normal building ventilation serve as part of the smoke control system, the expected performance of these systems in smoke control mode shall be addressed in the rational analysis required by Section 909.4 of the *International Building Code.*

607.2.2 Hazardous exhaust ducts. Fire dampers for hazardous exhaust duct systems shall comply with Section 510.

[BF] 607.3 Damper testing, ratings and actuation. Damper testing, ratings and actuation shall be in accordance with Sections 607.3.1 through 607.3.3.

[BF] 607.3.1 Damper testing. *Dampers* shall be listed and labeled in accordance with the standards in this section. *Fire dampers* shall comply with the requirements of UL 555. Only *fire dampers* and ceiling radiation dampers labeled for use in dynamic systems shall be installed in heating, ventilating and air-conditioning systems designed to operate with fans on during a fire. *Smoke dampers* shall comply with the requirements of UL 555S. *Combination fire/smoke dampers* shall comply with the requirements of both UL 555 and UL 555S. *Ceiling radiation dampers* shall comply with the requirements of UL 555C or shall be tested as part of a fire-resistance-rated floor/ceiling or roof/ceiling assembly in accordance with ASTM E 119 or UL 263. Corridor dampers shall comply with requirements of both UL 555 and UL 555S. Corridor dampers shall demonstrate acceptable closure performance when subjected to 150 feet per minute (0.76 mps) velocity across the face of the damper using the UL 555 fire exposure test.

[BF] 607.3.2 Damper rating. Damper ratings shall be in accordance with Sections 607.3.2.1 through 607.3.2.4.

[BF] 607.3.2.1 Fire damper ratings. Fire dampers shall have the minimum fire protection rating specified in Table 607.3.2.1 for the type of penetration.

[BF] TABLE 607.3.2.1
FIRE DAMPER RATING

TYPE OF PENETRATION	MINIMUM DAMPER RATING (hour)
Less than 3-hour fire-resistance-rated assemblies	$1^1/_2$
3-hour or greater fire-resistance-rated assemblies	3

[BF] 607.3.2.2 Smoke damper ratings. Smoke damper leakage ratings shall be Class I or II. Elevated temperature ratings shall be not less than 250°F (121°C).

[BF] 607.3.2.3 Combination fire/smoke damper ratings. Combination fire/smoke dampers shall have the minimum fire protection rating specified for fire dampers in Table 607.3.2.1 for the type of penetration and shall have a minimum smoke damper rating as specified in Section 607.3.2.2.

[BF] 607.3.2.4 Corridor damper ratings. Corridor dampers shall have the following minimum ratings.

1. One hour fire-resistance rating.

2. Class I or II leakage rating as specified in Section 607.3.2.2.

[BF] 607.3.3 Damper actuation. Damper actuation shall be in accordance with Sections 607.3.3.1 through 607.3.3.4 as applicable.

[BF] 607.3.3.1 Fire damper actuation device. The fire damper actuation device shall meet one of the following requirements:

1. The operating temperature shall be approximately 50°F (28°C) above the normal temperature within the duct system, but not less than 160°F (71°C).

2. The operating temperature shall be not more than 350°F (177°C) where located in a smoke control system complying with Section 909 of the *International Building Code.*

[BF] 607.3.3.2 Smoke damper actuation. The smoke damper shall close upon actuation of a *listed* smoke detector or detectors installed in accordance with Section 907.3 of the *International Building Code* and one of the following methods, as applicable:

1. Where a smoke damper is installed within a duct, a smoke detector shall be installed inside the duct or outside the duct with sampling tubes protruding into the duct. The detector or tubes within the duct shall be within 5 feet (1524 mm) of the damper. Air outlets and inlets shall not be located between the detector or tubes and the damper. The detector shall be *listed* for the air velocity, temperature and humidity anticipated at the point where it is installed. Other than in mechanical smoke control systems, dampers shall be closed upon fan shutdown where local smoke detectors require a minimum velocity to operate.

2. Where a smoke damper is installed above smoke barrier doors in a smoke barrier, a spot-type detector shall be installed on either side of the smoke barrier door opening. The detector shall be listed for releasing service if used for direct interface with the damper.

3. Where a smoke damper is installed within an unducted opening in a wall, a spot-type detector shall be installed within 5 feet (1524 mm) horizontally of the damper. The detector shall be listed for releasing service if used for direct interface with the damper.

4. Where a smoke damper is installed in a corridor wall or ceiling, the damper shall be permitted to be controlled by a smoke detection system installed in the corridor.

5. Where a smoke detection system is installed in all areas served by the duct in which the damper will

be located, the smoke dampers shall be permitted to be controlled by the smoke detection system.

[BF] 607.3.3.3 Combination fire/smoke damper actuation. Combination fire/smoke damper actuation shall be in accordance with Sections 607.3.3.1 and 607.3.3.2. Combination fire/smoke dampers installed in smoke control system shaft penetrations shall not be activated by local area smoke detection unless it is secondary to the smoke management system controls.

[BF] 607.3.3.4 Ceiling radiation damper actuation. The operating temperature of a ceiling radiation damper actuation device shall be 50°F (28°C) above the normal temperature within the duct system, but not less than 160°F (71°C).

[BF] 607.3.3.5 Corridor damper actuation. Corridor damper actuation shall be in accordance with Sections 607.3.3.1 and 607.3.3.2.

[BF] 607.4 Access and identification. Fire and smoke dampers shall be provided with an *approved* means of access, large enough to permit inspection and maintenance of the damper and its operating parts. The access shall not affect the integrity of fire-resistance-rated assemblies. The access openings shall not reduce the fire-resistance rating of the assembly. Access points shall be permanently identified on the exterior by a label having letters not less than 0.5 inch (12.7 mm) in height reading: FIRE/SMOKE DAMPER, SMOKE DAMPER or FIRE DAMPER. Access doors in ducts shall be tight fitting and suitable for the required duct construction.

[BF] 607.5 Where required. Fire dampers, smoke dampers, combination fire/smoke dampers, ceiling radiation dampers and corridor dampers shall be provided at the locations prescribed in Sections 607.5.1 through 607.5.7. Where an assembly is required to have both fire dampers and smoke dampers, combination fire/smoke dampers or a fire damper and smoke damper shall be provided.

[BF] 607.5.1 Fire walls. Ducts and air transfer openings permitted in fire walls in accordance with Section 706.11 of the *International Building Code* shall be protected with *listed* fire dampers installed in accordance with their listing.

[BF] 607.5.1.1 Horizontal exits. A *listed smoke damper* designed to resist the passage of smoke shall be provided at each point that a duct or air transfer opening penetrates a *fire wall* that serves as a horizontal *exit.*

[BF] 607.5.2 Fire barriers. Ducts and air transfer openings that penetrate fire barriers shall be protected with *listed* fire dampers installed in accordance with their listing. Ducts and air transfer openings shall not penetrate enclosures for interior exit stairways and ramps and exit passageways except as permitted by Sections 1023.5 and 1024.6, respectively, of the *International Building Code.*

Exception: Fire dampers are not required at penetrations of fire barriers where any of the following apply:

1. Penetrations are tested in accordance with ASTM E 119 or UL 263 as part of the fire-resistance-rated assembly.

2. Ducts are used as part of an *approved* smoke control system in accordance with Section 513 and where the fire damper would interfere with the operation of the smoke control system.

3. Such walls are penetrated by ducted HVAC systems, have a required fire-resistance rating of 1 hour or less, are in areas of other than Group H and are in buildings equipped throughout with an automatic sprinkler system in accordance with Section 903.3.1.1 or 903.3.1.2 of the *International Building Code.* For the purposes of this exception, a ducted HVAC system shall be a duct system for the structure's HVAC system. Such a duct system shall be constructed of sheet steel not less than 26 gage [0.0217 inch (0.55 mm)] thickness and shall be continuous from the air-handling *appliance* or *equipment* to the air outlet and inlet terminals.

[BF] 607.5.2.1 Horizontal exits. A *listed smoke damper* designed to resist the passage of smoke shall be provided at each point that a duct or air transfer opening penetrates a *fire barrier* that serves as a horizontal *exit.*

[BF] 607.5.3 Fire partitions. Ducts and air transfer openings that penetrate fire partitions shall be protected with *listed* fire dampers installed in accordance with their listing.

Exception: In occupancies other than Group H, fire dampers are not required where any of the following apply:

1. Corridor walls in buildings equipped throughout with an automatic sprinkler system in accordance with Section 903.3.1.1 or 903.3.1.2 of the *International Building Code* and the duct is protected as a through penetration in accordance with Section 714 of the *International Building Code.*

2. The partitions are tenant partitions in covered and open mall buildings where the walls are not required by provisions elsewhere in the *International Building Code* to extend to the underside of the floor or roof sheathing, slab or deck above.

3. The duct system is constructed of *approved* materials in accordance with Section 603 and the duct penetrating the wall complies with all of the following requirements:

   3.1. The duct shall not exceed 100 square inches (0.06 m²).

   3.2. The duct shall be constructed of steel not less than 0.0217 inch (0.55 mm) in thickness.

   3.3. The duct shall not have openings that communicate the corridor with adjacent spaces or rooms.

   3.4. The duct shall be installed above a ceiling.

   3.5. The duct shall not terminate at a wall register in the fire-resistance-rated wall.

3.6. A minimum 12-inch-long (305 mm) by 0.060-inch-thick (1.52 mm) steel sleeve shall be centered in each duct opening. The sleeve shall be secured to both sides of the wall and all four sides of the sleeve with minimum $1^1/_2$-inch by $1^1/_2$-inch by 0.060-inch (38 mm by 38 mm by 1.52 mm) steel retaining angles. The retaining angles shall be secured to the sleeve and the wall with No. 10 (M5) screws. The annular space between the steel sleeve and the wall opening shall be filled with rock (mineral) wool batting on all sides.

4. Such walls are penetrated by ducted HVAC systems, have a required fire-resistance rating of 1 hour or less, and are in areas of other than Group H and are in buildings equipped throughout with an automatic sprinkler system in accordance with Section 903.3.1.1 or 903.3.1.2 of the *International Building Code*. For the purposes of this exception, a ducted HVAC system shall be a duct system for conveying supply, return or exhaust air as part of the structure's HVAC system. Such a duct system shall be constructed of sheet steel not less than 26 gage in thickness and shall be continuous from the air-handling appliance or equipment to the air outlet and inlet terminals.

[BF] 607.5.4 Corridors/smoke barriers. A *listed* smoke damper designed to resist the passage of smoke shall be provided at each point a duct or air transfer opening penetrates a smoke barrier wall or a corridor enclosure required to have smoke and draft control doors in accordance with the *International Building Code*.

A corridor damper shall be provided where corridor ceilings, constructed as required for the corridor walls as permitted in Section 708.4, Exception 3, of the *International Building Code*, are penetrated.

A ceiling radiation damper shall be provided where the ceiling membrane of a fire-resistance-rated floor/ceiling or roof/ceiling assembly, constructed as permitted in Section 708.4, Exception 2, of the *International Building Code*, is penetrated.

Smoke dampers and smoke damper actuation methods shall comply with Section 607.5.4.1.

Exceptions:

1. Smoke dampers are not required in corridor penetrations where the building is equipped throughout with an *approved* smoke control system in accordance with Section 513 and smoke dampers are not necessary for the operation and control of the system.

2. Smoke dampers are not required in smoke barrier penetrations where the openings in ducts are limited to a single smoke compartment and the ducts are constructed of steel.

3. Smoke dampers are not required in corridor penetrations where the duct is constructed of steel not less than 0.019 inch (0.48 mm) in thickness and there are no openings serving the corridor.

4. Smoke dampers are not required in smoke barriers required by Section 407.5 of the *International Building Code* for Group I-2 Condition 2 where the HVAC system is fully ducted in accordance with Section 603 and where buildings are equipped throughout with an automatic sprinkler system in accordance with Section 903.3.1.1 of the *International Building Code* and equipped with quick-response sprinklers in accordance with Section 903.3.2 of the *International Building Code*.

[BF] 607.5.4.1 Smoke damper. Smoke dampers shall close as required by Section 607.3.3.2.

[BF] 607.5.5 Shaft enclosures. Shaft enclosures that are permitted to be penetrated by ducts and air transfer openings shall be protected with *approved* fire and smoke dampers installed in accordance with their listing.

Exceptions:

1. Fire dampers are not required at penetrations of shafts where any of the following apply:

    1.1. Steel exhaust subducts extend not less than 22 inches (559 mm) vertically in exhaust shafts provided that there is a continuous airflow upward to the outdoors.

    1.2. Penetrations are tested in accordance with ASTM E 119 or UL 263 as part of the fire-resistance-rated assembly.

    1.3. Ducts are used as part of an *approved* smoke control system in accordance with Section 909 of the *International Building Code*, and where the fire damper will interfere with the operation of the smoke control system.

    1.4. The penetrations are in parking garage exhaust or supply shafts that are separated from other building shafts by not less than 2-hour fire-resistance-rated construction.

2. In Group B and R occupancies equipped throughout with an automatic sprinkler system in accordance with Section 903.3.1.1 of the *International Building Code*, smoke dampers are not required at penetrations of shafts where kitchen, clothes dryer, bathroom and toilet room exhaust openings with steel exhaust subducts, having a minimum thickness of 0.0187 inch (0.4712 mm) (No. 26 gage), extend not less than 22 inches (559 mm) vertically and the exhaust fan at the upper terminus is powered continuously in accordance with the provisions of Section 909.11 of the *International Building Code*, and maintains airflow upward to the outdoors.

3. Smoke dampers are not required at penetrations of exhaust or supply shafts in parking garages

that are separated from other building shafts by not less than 2-hour fire-resistance-rated construction.

4. Smoke dampers are not required at penetrations of shafts where ducts are used as part of an *approved* mechanical smoke control system designed in accordance with Section 909 of the *International Building Code* and where the smoke damper will interfere with the operation of the smoke control system.

5. Fire dampers and combination fire/smoke dampers are not required in kitchen and clothes dryer exhaust systems installed in accordance with this code.

[BF] 607.5.5.1 Enclosure at the bottom. Shaft enclosures that do not extend to the bottom of the building or structure shall be protected in accordance with Section 713.11 of the *International Building Code*.

[BF] 607.5.6 Exterior walls. Ducts and air transfer openings in fire-resistance-rated exterior walls required to have protected openings in accordance with Section 705.10 of the *International Building Code* shall be protected with *listed* fire dampers installed in accordance with their listing.

[BF] 607.5.7 Smoke partitions. A *listed* smoke damper designed to resist the passage of smoke shall be provided at each point where an air transfer opening penetrates a smoke partition. Smoke dampers and smoke damper actuation methods shall comply with Section 607.3.3.2.

Exception: Where the installation of a smoke damper will interfere with the operation of a required smoke control system in accordance with Section 513, *approved* alternate protection shall be used.

[BF] 607.6 Horizontal assemblies. Penetrations by air ducts of a floor, floor/ceiling assembly or the ceiling membrane of a roof/ceiling assembly shall be protected by a shaft enclosure that complies with Section 713 and Sections 717.6.1 through 717.6.3 of the *International Building Code* or shall comply with Sections 607.6.1 through 607.6.3.

[BF] 607.6.1 Through penetrations. In occupancies other than Groups I-2 and I-3, a duct constructed of *approved* materials in accordance with Section 603 that penetrates a fire-resistance-rated floor/ceiling assembly that connects not more than two stories is permitted without shaft enclosure protection provided that a *listed* fire damper is installed at the floor line or the duct is protected in accordance with Section 714.4 of the *International Building Code*. For air transfer openings, see Item 6, Section 712.1.9 of the *International Building Code*.

Exception: A duct is permitted to penetrate three floors or less without a fire damper at each floor provided it meets all of the following requirements.

1. The duct shall be contained and located within the cavity of a wall and shall be constructed of steel having a minimum thickness of 0.0187 inch (0.4712 mm) (No. 26 gage).

2. The duct shall open into only one *dwelling unit* or *sleeping unit* and the duct system shall be continuous from the unit to the exterior of the building.

3. The duct shall not exceed a 4-inch (102 mm) nominal diameter and the total area of such ducts shall not exceed 100 square inches for any 100 square feet (64 516 mm^2 per 9.3 m^2) of the floor area.

4. The annular space around the duct is protected with materials that prevent the passage of flame and hot gases sufficient to ignite cotton waste when subjected to ASTM E 119 or UL 263 time-temperature conditions under a minimum positive pressure differential of 0.01 inch (2.49 Pa) of water at the location of the penetration for the time period equivalent to the fire-resistance rating of the construction penetrated.

5. Grille openings located in a ceiling of a fire-resistance-rated floor/ceiling or roof/ceiling assembly shall be protected with a *listed* ceiling radiation damper installed in accordance with Section 607.6.2.1.

[BF] 607.6.2 Membrane penetrations. Ducts and air transfer openings constructed of *approved* materials, in accordance with Section 603, that penetrate the ceiling membrane of a fire-resistance-rated floor/ceiling or roof/ceiling assembly shall be protected with one of the following:

1. A shaft enclosure in accordance with Section 713 of the *International Building Code*.

2. A *listed* ceiling radiation damper installed at the ceiling line where a duct penetrates the ceiling of a fire-resistance-rated floor/ceiling or roof/ceiling assembly.

3. A *listed* ceiling radiation damper installed at the ceiling line where a diffuser with no duct attached penetrates the ceiling of a fire-resistance-rated floor/ceiling or roof/ceiling assembly.

[BF] 607.6.2.1 Ceiling radiation dampers. *Ceiling radiation dampers* shall be tested in accordance with Section 607.3.1. *Ceiling radiation dampers* shall be installed in accordance with the details listed in the fire-resistance-rated assembly and the manufacturer's installation instructions and the listing. *Ceiling radiation dampers* are not required where any of the following apply:

1. Tests in accordance with ASTM E 119 or UL 263 have shown that ceiling radiation dampers are not necessary to maintain the fire-resistance rating of the assembly.

2. Where exhaust duct penetrations are protected in accordance with Section 714.4.1.2 of the *International Building Code*, are located within the cavity of a wall, and do not pass through another dwelling unit or tenant space.

3. Where duct and air transfer openings are protected with a duct outlet protection system tested as part of a fire-resistance-rated assembly in accordance with ASTM E 119 or UL 263.

[BF] 607.6.3 Nonfire-resistance-rated floor assemblies. Duct systems constructed of approved materials in accordance with Section 603 that penetrate nonfire-resistance-rated floor assemblies shall be protected by any of the following methods:

1. A shaft enclosure in accordance with Section 713 of the *International Building Code*.

2. The duct connects not more than two stories, and the annular space around the penetrating duct is protected with an *approved* noncombustible material that resists the free passage of flame and the products of *combustion*.

3. In floor assemblies composed of noncombustible materials, a shaft shall not be required where the duct connects not more than three stories, and the annular space around the penetrating duct is protected with an approved noncombustible material that resists the free passage of flame and the products of combustion and a fire damper is installed at each floor line.

   Exception: Fire dampers are not required in ducts within individual residential *dwelling units*.

[BF] 607.7 Flexible ducts and air connectors. Flexible ducts and air connectors shall not pass through any fire-resistance-rated assembly.

CHAPTER 7

COMBUSTION AIR

SECTION 701
GENERAL

701.1 Scope. Solid fuel-burning *appliances* shall be provided with *combustion air* in accordance with the appliance manufacturer's installation instructions. Oil-fired *appliances* shall be provided with *combustion air* in accordance with NFPA 31. The methods of providing *combustion air* in this chapter do not apply to fireplaces, fireplace stoves and direct-vent *appliances*. The requirements for combustion and dilution air for gas-fired *appliances* shall be in accordance with the *International Fuel Gas Code*.

701.2 Dampered openings. Where combustion air openings are provided with volume, smoke or fire dampers, the dampers shall be interlocked with the firing cycle of the appliances served, so as to prevent operation of any appliance that draws combustion air from the room or space when any of the dampers are closed. Manual dampers shall not be installed in combustion air ducts. Ducts not provided with dampers and that pass through rated construction shall be enclosed in a shaft in accordance with the *International Building Code*.

CHAPTER 8

CHIMNEYS AND VENTS

SECTION 801
GENERAL

801.1 Scope. This chapter shall govern the installation, maintenance, repair and approval of factory-built chimneys, *chimney* liners, vents and connectors. This chapter shall govern the utilization of masonry chimneys. Gas-fired *appliances* shall be vented in accordance with the *International Fuel Gas Code*.

801.2 General. Every fuel-burning *appliance* shall discharge the products of *combustion* to a vent, factory-built *chimney* or masonry *chimney*, except for *appliances* vented in accordance with Section 804. The *chimney* or vent shall be designed for the type of *appliance* being vented.

> **Exception:** Commercial cooking *appliances* vented by a Type I hood installed in accordance with Section 507.

> **801.2.1 Oil-fired appliances.** Oil-fired *appliances* shall be vented in accordance with this code and NFPA 31.

801.3 Masonry chimneys. Masonry *chimneys* shall be constructed in accordance with the *International Building Code*.

801.4 Positive flow. Venting systems shall be designed and constructed so as to develop a positive flow adequate to convey all *combustion* products to the outside atmosphere.

801.5 Design. Venting systems shall be designed in accordance with this chapter or shall be *approved* engineered systems.

801.6 Minimum size of chimney or vent. Except as otherwise provided for in this chapter, the size of the *chimney* or vent, serving a single *appliance*, except engineered systems, shall have a minimum area equal to the area of the *appliance* connection.

801.7 Solid fuel appliance flues. The cross-sectional area of a flue serving a solid-fuel-burning *appliance* shall be not greater than three times the cross-sectional area of the *appliance* flue collar or flue outlet.

801.8 Abandoned inlet openings. Abandoned inlet openings in chimneys and vents shall be closed by an *approved* method.

801.9 Positive pressure. Where an *appliance* equipped with a forced or induced draft system creates a positive pressure in the venting system, the venting system shall be designed and *listed* for positive pressure applications.

801.10 Connection to fireplace. Connection of *appliances* to *chimney* flues serving fireplaces shall be in accordance with Sections 801.10.1 through 801.10.3.

> **801.10.1 Closure and access.** A noncombustible seal shall be provided below the point of connection to prevent entry of room air into the flue. Means shall be provided for *access* to the flue for inspection and cleaning.

> **801.10.2 Connection to factory-built fireplace flue.** An *appliance* shall not be connected to a flue serving a factory-built fireplace unless the *appliance* is specifically *listed* for such installation. The connection shall be made in accordance with the *appliance* manufacturer's installation instructions.

> **801.10.3 Connection to masonry fireplace flue.** A connector shall extend from the *appliance* to the flue serving a masonry fireplace such that the flue gases are exhausted directly into the flue. The connector shall be provided with access or shall be removable for inspection and cleaning of both the connector and the flue. *Listed* direct connection devices shall be installed in accordance with their listing.

801.11 Multiple solid fuel prohibited. A solid fuel-burning *appliance* or fireplace shall not connect to a *chimney* passageway venting another *appliance*.

801.12 Chimney entrance. Connectors shall connect to a *chimney* flue at a point not less than 12 inches (305 mm) above the lowest portion of the interior of the *chimney* flue.

801.13 Cleanouts. Masonry *chimney* flues shall be provided with a cleanout opening having a minimum height of 6 inches (152 mm). The upper edge of the opening shall be located not less than 6 inches (152 mm) below the lowest *chimney* inlet opening. The cleanout shall be provided with a tight-fitting, noncombustible cover.

> **Exception:** Cleanouts shall not be required for *chimney* flues serving masonry fireplaces, if such flues are provided with access through the fireplace opening.

801.14 Connections to exhauster. *Appliance* connections to a *chimney* or vent equipped with a power exhauster shall be made on the inlet side of the exhauster. Joints and piping on the positive pressure side of the exhauster shall be *listed* for positive pressure applications as specified by the manufacturer's installation instructions for the exhauster.

801.15 Fuel-fired appliances. Masonry chimneys utilized to vent fuel-fired *appliances* shall be located, constructed and sized as specified in the manufacturer's installation instructions for the *appliances* being vented.

801.16 Flue lining. Masonry chimneys shall be lined. The lining material shall be compatible with the type of *appliance* connected, in accordance with the *appliance* listing and manufacturer's installation instructions. *Listed* materials used as flue linings shall be installed in accordance with their listings and the manufacturer's instructions.

> **801.16.1 Residential and low-heat appliances (general).** Flue lining systems for use with residential-type and low-heat appliances shall be limited to the following:
>
> 1. Clay flue lining complying with the requirements of ASTM C 315 or equivalent. Clay flue lining shall be installed in accordance with the *International Building Code*.
>
> 2. *Listed* and *labeled* chimney lining systems complying with UL 1777.

3. Other *approved* materials that will resist, without cracking, softening or corrosion, flue gases and condensate at temperatures up to 1,800°F (982°C).

801.17 Space around lining. The space surrounding a flue lining system or other vent installed within a masonry *chimney* shall not be used to vent any other *appliance*. This shall not prevent the installation of a separate flue lining in accordance with the manufacturer's installation instructions and this code.

801.18 Existing chimneys and vents. Where an *appliance* is permanently disconnected from an existing *chimney* or vent, or where an *appliance* is connected to an existing *chimney* or vent during the process of a new installation, the *chimney* or vent shall comply with Sections 801.18.1 through 801.18.4.

801.18.1 Size. The *chimney* or vent shall be resized as necessary to control flue gas condensation in the interior of the *chimney* or vent and to provide the *appliance* or *appliances* served with the required draft. For the venting of oil-fired *appliances* to masonry chimneys, the resizing shall be in accordance with NFPA 31.

801.18.2 Flue passageways. The flue gas passageway shall be free of obstructions and combustible deposits and shall be cleaned if previously used for venting a solid or liquid fuel-burning *appliance* or fireplace. The flue liner, *chimney* inner wall or vent inner wall shall be continuous and shall be free of cracks, gaps, perforations or other damage or deterioration which would allow the escape of *combustion* products, including gases, moisture and creosote. Where an oil-fired *appliance* is connected to an existing masonry *chimney*, such *chimney* flue shall be repaired or relined in accordance with NFPA 31.

801.18.3 Cleanout. Masonry chimneys shall be provided with a cleanout opening complying with Section 801.13.

801.18.4 Clearances. Chimneys and vents shall have airspace *clearance* to combustibles in accordance with the *International Building Code* and the *chimney* or vent manufacturer's installation instructions.

Exception: Masonry chimneys without the required airspace *clearances* shall be permitted to be used if lined or relined with a *chimney* lining system *listed* for use in chimneys with reduced *clearances* in accordance with UL 1777. The *chimney clearance* shall be not less than permitted by the terms of the *chimney* liner listing and the manufacturer's instructions.

801.18.4.1 Fireblocking. Noncombustible fireblocking shall be provided in accordance with the *International Building Code*.

801.19 Multistory prohibited. Common venting systems for appliances located on more than one floor level shall be prohibited, except where all of the appliances served by the common vent are located in rooms or spaces that are accessed only from the outdoors. The *appliance* enclosures shall not communicate with the occupiable areas of the building.

801.20 Plastic vent joints. Plastic pipe and fittings used to vent appliances shall be installed in accordance with the *appliance* manufacturer's installation instructions.

SECTION 802
VENTS

802.1 General. Vent systems shall be *listed* and *labeled*. Type L vents and pellet vents shall be tested in accordance with UL 641.

802.2 Vent application. The application of vents shall be in accordance with Table 802.2.

TABLE 802.2
VENT APPLICATION

VENT TYPES	APPLIANCE TYPES
Type L oil vents	Oil-burning appliances listed and labeled for venting with Type L vents; gas appliances listed and labeled for venting with Type B vents.
Pellet vents	Pellet fuel-burning appliances listed and labeled for venting with pellet vents.

802.3 Installation. Vent systems shall be sized, installed and terminated in accordance with the vent and *appliance* manufacturer's installation instructions.

802.4 Vent termination caps required. Type L vents shall terminate with a *listed* and *labeled* cap in accordance with the vent manufacturer's installation instructions.

802.5 Type L vent terminations. Type L vents shall terminate not less than 2 feet (610 mm) above the highest point of the roof penetration and not less than 2 feet (610 mm) higher than any portion of a building within 10 feet (3048 mm).

802.6 Minimum vent heights. Vents shall terminate not less than 5 feet (1524 mm) in vertical height above the highest connected *appliance* flue collar.

Exceptions:

1. Venting systems of direct vent *appliances* shall be installed in accordance with the *appliance* and the vent manufacturer's instructions.

2. Appliances *listed* for outdoor installations incorporating integral venting means shall be installed in accordance with their listings and the manufacturer's installation instructions.

3. Pellet vents shall be installed in accordance with the *appliance* and the vent manufacturer's installation instructions.

802.7 Support of vents. All portions of vents shall be adequately supported for the design and weight of the materials employed.

802.8 Insulation shield. Where vents pass through insulated assemblies, an insulation shield constructed of not less than No. 26 gage sheet metal shall be installed to provide *clearance* between the vent and the insulation material. The *clearance* shall be not less than the *clearance* to combustibles specified by the vent manufacturer's installation instructions. Where vents pass through attic space, the shield shall terminate not less than 2 inches (51 mm) above the insulation materials and shall be secured in place to prevent displacement. Insulation shields provided as part of a *listed* vent system shall be installed in accordance with the manufacturer's installation instructions.

802.9 Door swing. Appliance and equipment vent terminals shall be located such that doors cannot swing within 12 inches (305 mm) horizontally of the vent terminals. Door-stops or closers shall not be installed to obtain this clearance.

SECTION 803
CONNECTORS

803.1 Connectors required. Connectors shall be used to connect *appliances* to the vertical *chimney* or vent, except where the *chimney* or vent is attached directly to the *appliance*.

803.2 Location. Connectors shall be located entirely within the room in which the connecting *appliance* is located, except as provided for in Section 803.10.4. Where passing through an unheated space, a connector shall not be constructed of single-wall pipe.

803.3 Size. The connector shall not be smaller than the size of the flue collar supplied by the manufacturer of the *appliance*. Where the *appliance* has more than one flue outlet, and in the absence of the manufacturer's specific instructions, the connector area shall be not less than the combined area of the flue outlets for which it acts as a common connector.

803.4 Branch connections. Branch connections to the vent connector shall be made in accordance with the vent manufacturer's instructions.

803.5 Manual dampers. Manual dampers shall not be installed in connectors except in *chimney* connectors serving solid fuel-burning *appliances*.

803.6 Automatic dampers. Automatic dampers shall be *listed* and *labeled* in accordance with UL 17 for oil-fired heating appliances. The dampers shall be installed in accordance with the manufacturer's instructions. An automatic vent damper device shall not be installed on an existing *appliance* unless the *appliance* is *listed* and *labeled* and the device is installed in accordance with the terms of its listing. The name of the installer and date of installation shall be marked on a label affixed to the damper device.

803.7 Connectors serving two or more appliances. Where two or more connectors enter a common vent or *chimney*, the smaller connector shall enter at the highest level consistent with available headroom or *clearance* to combustible material.

803.8 Vent connector construction. Vent connectors shall be constructed of metal. The minimum thickness of the connector shall be 0.0136 inch (0.345 mm) (No. 28 gage) for galvanized steel, 0.022 inch (0.6 mm) (No. 26 B & S gage) for copper, and 0.020 inch (0.5 mm) (No. 24 B & S gage) for aluminum.

803.9 Chimney connector construction. *Chimney* connectors for low-heat *appliances* shall be of sheet steel pipe having resistance to corrosion and heat not less than that of galvanized steel specified in Table 803.9(1). Connectors for medium-heat *appliances* and high-heat appliances shall be of sheet steel not less than the thickness specified in Table 803.9(2).

TABLE 803.9(1)
MINIMUM CHIMNEY CONNECTOR
THICKNESS FOR LOW-HEAT APPLIANCES

DIAMETER OF CONNECTOR (inches)	MINIMUM NOMINAL THICKNESS (galvanized) (inches)
5 and smaller	0.022 (No. 26 gage)
Larger than 5 and up to 10	0.028 (No. 24 gage)
Larger than 10 and up to 16	0.034 (No. 22 gage)
Larger than 16	0.064 (No. 16 gage)

For SI: 1 inch = 25.4 mm.

TABLE 803.9(2)
MINIMUM CHIMNEY CONNECTOR
THICKNESS FOR MEDIUM- AND HIGH-HEAT APPLIANCES

AREA (square inches)	EQUIVALENT ROUND DIAMETER (inches)	MINIMUM THICKNESS (inches)
0-154	0-14	0.0575 (No. 16 gage)
155-201	15-16	0.075 (No. 14 gage)
202-254	17-18	0.0994 (No. 12 gage)
Greater than 254	Greater than 18	0.1292 (No. 10 gage)

For SI: 1 inch = 25.4 mm, 1 square inch = 645.16 mm².

803.10 Installation. Connectors shall be installed in accordance with Sections 803.10.1 through 803.10.6.

803.10.1 Supports and joints. Connectors shall be supported in an *approved* manner, and joints shall be fastened with sheet metal screws, rivets or other *approved* means.

803.10.2 Length. The maximum horizontal length of a single-wall connector shall be 75 percent of the height of the *chimney* or vent.

803.10.3 Connection. The connector shall extend to the inner face of the *chimney* or vent liner, but not beyond. A connector entering a masonry *chimney* shall be cemented to masonry in an *approved* manner. Where thimbles are installed to facilitate removal of the connector from the masonry *chimney*, the thimble shall be permanently cemented in place with high-temperature cement.

803.10.4 Connector pass-through. *Chimney* connectors shall not pass through any floor or ceiling, nor through a fire-resistance-rated wall assembly. *Chimney* connectors for domestic-type *appliances* shall not pass through walls or partitions constructed of combustible material to reach a masonry *chimney* except where one of the following apply:

1. The connector is *labeled* for wall pass-through and is installed in accordance with the manufacturer's instructions.

2. The connector is put through a device *labeled* for wall pass-through.

3. The connector has a diameter not larger than 10 inches (254 mm) and is installed in accordance with one of the methods in Table 803.10.4. Concealed metal parts of the pass-through system in contact

with flue gases shall be of stainless steel or equivalent material that resists corrosion, softening or cracking up to 1,800°F (980°C).

803.10.5 Pitch. Connectors shall rise vertically to the *chimney* or vent with a minimum pitch equal to one-fourth unit vertical in 12 units horizontal (2-percent slope).

TABLE 803.10.4
CHIMNEY CONNECTOR SYSTEMS AND CLEARANCES
TO COMBUSTIBLE WALL MATERIALS FOR
DOMESTIC HEATING APPLIANCES[a, b, c, d]

System A (12-inch clearance)	A 3.5-inch-thick brick wall shall be framed into the combustible wall. An 0.625-inch-thick fire-clay liner (ASTM C 315 or equivalent)[e] shall be firmly cemented in the center of the brick wall maintaining a 12-inch clearance to combustibles. The clay liner shall run from the outer surface of the bricks to the inner surface of the chimney liner.
System B (9-inch clearance)	A labeled solid-insulated factory-built chimney section (1-inch insulation) the same inside diameter as the connector shall be utilized. Sheet steel supports cut to maintain a 9-inch clearance to combustibles shall be fastened to the wall surface and to the chimney section. Fasteners shall not penetrate the chimney flue liner. The chimney length shall be flush with the masonry chimney liner and sealed to the masonry with water-insoluble refractory cement. Chimney manufacturers' parts shall be utilized to securely fasten the chimney connector to the chimney section.
System C (6-inch clearance)	A steel ventilated thimble having a minimum thickness of 0.0236 inch (No. 24 gage) having two 1-inch air channels shall be installed with a steel chimney connector. Steel supports shall be cut to maintain a 6-inch clearance between the thimble and combustibles. The chimney connector and steel supports shall have a minimum thickness of 0.0236 inch (No. 24 gage). One side of the support shall be fastened to the wall on all sides. Glass-fiber insulation shall fill the 6-inch space between the thimble and the supports.
System D (2-inch clearance)	A labeled solid-insulated factory-built chimney section (1-inch insulation) with a diameter 2 inches larger than the chimney connector shall be installed with a steel chimney connector having a minimum thickness of 0.0236 inch (24 gage). Sheet steel supports shall be positioned to maintain a 2-inch clearance to combustibles and to hold the chimney connector to ensure that a 1-inch airspace surrounds the chimney connector through the chimney section. The steel support shall be fastened to the wall on all sides and the chimney section shall be fastened to the supports. Fasteners shall not penetrate the liner of the chimney section.

For SI: 1 inch = 25.4 mm, 1.0 Btu x in/ft² • h • °F = 0.144 W/m² • K.

a. Insulation material that is part of the wall pass-through system shall be noncombustible and shall have a thermal conductivity of 1.0 Btu x in/ft² • h • °F or less.

b. All clearances and thicknesses are minimums.

c. Materials utilized to seal penetrations for the connector shall be noncombustible.

d. Connectors for all systems except System B shall extend through the wall pass-through system to the inner face of the flue liner.

e. ASTM C 315.

803.10.6 Clearances. Connectors shall have a minimum *clearance* to combustibles in accordance with Table 803.10.6. The clearances specified in Table 803.10.6 apply, except where the listing and labeling of an *appliance* specifies a different *clearance*, in which case the *labeled clearance* shall apply. The *clearance* to combustibles for connectors shall be reduced only in accordance with Section 308.

TABLE 803.10.6
CONNECTOR CLEARANCES TO COMBUSTIBLES

TYPE OF APPLIANCE	MINIMUM CLEARANCE (inches)
Domestic-type appliances	
Chimney and vent connectors Electric and oil incinerators Oil and solid-fuel appliances Oil appliances labeled for venting with Type L vents	18 18 9
Commercial, industrial-type appliances	
Low-heat appliances Chimney connectors Oil and solid-fuel boilers, furnace and water heaters Oil unit heaters Other low-heat industrial appliances	18 18 18
Medium-heat appliances Chimney connectors All oil and solid-fuel appliances	36
High-heat appliances Masonry or metal connectors All oil and solid-fuel appliances	(As determined by the code official)

For SI: 1 inch = 25.4 mm.

SECTION 804
DIRECT-VENT, INTEGRAL VENT AND MECHANICAL DRAFT SYSTEMS

804.1 Direct-vent terminations. Vent terminals for *direct-vent appliances* shall be installed in accordance with the manufacturer's instructions.

804.2 Appliances with integral vents. *Appliances* incorporating integral venting means shall be installed in accordance with their listings and the manufacturer's installation instructions.

804.2.1 Terminal clearances. *Appliances* designed for natural draft venting and incorporating integral venting means shall be located so that a minimum *clearance* of 9 inches (229 mm) is maintained between vent terminals and from any openings through which *combustion* products enter the building. *Appliances* using forced draft venting shall be located so that a minimum clearance of 12 inches (305 mm) is maintained between vent terminals and from any openings through which *combustion* products enter the building.

804.3 Mechanical draft systems. Mechanical draft systems of either forced or induced draft design shall be listed and labeled in accordance with UL 378 and shall comply with Sections 804.3.1 through 804.3.7.

804.3.1 Forced draft systems. Forced draft systems and all portions of induced draft systems under positive pressure during operation shall be designed and installed so as to be gas tight to prevent leakage of *combustion* products into a building.

804.3.2 Automatic shutoff. Power exhausters serving automatically fired *appliances* shall be electrically connected to each *appliance* to prevent operation of the *appliance* when the power exhauster is not in operation.

804.3.3 Termination. The termination of *chimneys* or vents equipped with power exhausters shall be located not less than 10 feet (3048 mm) from the lot line or from adjacent buildings. The exhaust shall be directed away from the building.

804.3.4 Horizontal terminations. Horizontal terminations shall comply with the following requirements:

1. Where located adjacent to walkways, the termination of mechanical draft systems shall be not less than 7 feet (2134 mm) above the level of the walkway.

2. Vents shall terminate at least 3 feet (914 mm) above any forced air inlet located within 10 feet (3048 mm).

3. The vent system shall terminate at least 4 feet (1219 mm) below, 4 feet (1219 mm) horizontally from or 1 foot (305 mm) above any door, window or gravity air inlet into the building.

4. The vent termination point shall not be located closer than 3 feet (914 mm) to an interior corner formed by two walls perpendicular to each other.

5. The vent termination shall not be mounted directly above or within 3 feet (914 mm) horizontally from an oil tank vent or gas meter.

6. The bottom of the vent termination shall be located at least 12 inches (305 mm) above finished grade.

804.3.5 Vertical terminations. Vertical terminations shall comply with the following requirements:

1. Where located adjacent to walkways, the termination of mechanical draft systems shall be not less than 7 feet (2134 mm) above the level of the walkway.

2. Vents shall terminate not less than 3 feet (914 mm) above any forced air inlet located within 10 feet (3048 mm) horizontally.

3. Where the vent termination is located below an adjacent roof structure, the termination point shall be located not less than 3 feet (914 mm) from such structure.

4. The vent shall terminate not less than 4 feet (1219 mm) below, 4 feet (1219 mm) horizontally from or 1 foot (305 mm) above any door, window or gravity air inlet for the building.

5. A vent cap shall be installed to prevent rain from entering the vent system.

6. The vent termination shall be located not less than 3 feet (914 mm) horizontally from any portion of the roof structure.

804.3.6 Exhauster connections. An *appliance* vented by natural draft shall not be connected into a vent, *chimney* or vent connector on the discharge side of a mechanical flue exhauster.

804.3.7 Exhauster sizing. Mechanical flue exhausters and the vent system served shall be sized and installed in accordance with the manufacturer's installation instructions.

804.3.8 Mechanical draft systems for manually fired appliances and fireplaces. A mechanical draft system shall be permitted to be used with manually fired appliances and fireplaces where such system complies with all of the following requirements:

1. The mechanical draft device shall be listed and labeled in accordance with UL 378, and shall be installed in accordance with the manufacturer's instructions.

2. A device shall be installed that produces visible and audible warning upon failure of the mechanical draft device or loss of electrical power, at any time that the mechanical draft device is turned on. This device shall be equipped with a battery backup if it receives power from the building wiring.

3. A smoke detector shall be installed in the room with the *appliance* or fireplace. This device shall be equipped with a battery backup if it receives power from the building wiring.

SECTION 805
FACTORY-BUILT CHIMNEYS

805.1 Listing. Factory-built *chimneys* shall be *listed* and *labeled* and shall be installed and terminated in accordance with the manufacturer's installation instructions.

805.2 Solid fuel appliances. Factory-built *chimneys* installed in *dwelling units* with solid fuel-burning appliances shall comply with the Type HT requirements of UL 103 and shall be marked "Type HT" and "Residential Type and Building Heating *Appliance Chimney.*"

Exception: *Chimneys* for use with open *combustion* chamber fireplaces shall comply with the requirements of UL 103 and shall be marked "Residential Type and Building Heating *Appliance Chimney.*"

Chimneys for use with open *combustion* chamber appliances installed in buildings other than *dwelling units* shall comply with the requirements of UL 103 and shall be marked "Building Heating *Appliance Chimney*" or "Residential Type and Building Heating *Appliance Chimney.*"

805.3 Factory-built chimney offsets. Where a factory-built chimney assembly incorporates offsets, no part of the chimney shall be at an angle of more than 30 degrees (0.52 rad) from vertical at any point in the assembly and the chimney assembly shall not include more than four elbows.

805.4 Support. Where factory-built *chimneys* are supported by structural members, such as joists and rafters, such members shall be designed to support the additional load.

805.5 Medium-heat appliances. Factory-built *chimneys* for medium-heat appliances producing flue gases having a temperature above 1,000°F (538°C) measured at the entrance to the *chimney* shall comply with UL 959.

805.6 Decorative shrouds. Decorative shrouds shall not be installed at the termination of factory-built *chimneys* except where such shrouds are *listed* and *labeled* for use with the specific factory-built *chimney* system and are installed in accordance with Section 304.1.

805.7 Factory-built fireplaces. Chimneys for use with factory-built fireplaces shall comply with the requirements of UL 127.

SECTION 806
METAL CHIMNEYS

806.1 General. Metal *chimneys* shall be constructed and installed in accordance with NFPA 211.

CHAPTER 9

SPECIFIC APPLIANCES, FIREPLACES AND SOLID FUEL-BURNING EQUIPMENT

SECTION 901
GENERAL

901.1 Scope. This chapter shall govern the approval, design, installation, construction, maintenance, *alteration* and repair of the appliances and *equipment* specifically identified herein and factory-built fireplaces. The approval, design, installation, construction, maintenance, *alteration* and repair of gas-fired appliances shall be regulated by the *International Fuel Gas Code*.

901.2 General. The requirements of this chapter shall apply to the mechanical *equipment* and appliances regulated by this chapter, in addition to the other requirements of this code.

901.3 Hazardous locations. Fireplaces and solid fuel-burning appliances shall not be installed in hazardous locations.

SECTION 902
MASONRY FIREPLACES

902.1 General. Masonry fireplaces shall be constructed in accordance with the *International Building Code*.

902.2 Fireplace accessories. Listed and labeled fireplace accessories shall be installed in accordance with the conditions of the listing and the manufacturer's instructions. Fireplace accessories shall comply with UL 907.

SECTION 903
FACTORY-BUILT FIREPLACES

903.1 General. Factory-built fireplaces shall be *listed* and *labeled* and shall be installed in accordance with the conditions of the listing. Factory-built fireplaces shall be tested in accordance with UL 127.

903.2 Hearth extensions. Hearth extensions of approved factory-built fireplaces shall be installed in accordance with the listing of the fireplace. The hearth extension shall be readily distinguishable from the surrounding floor area. Listed and labeled hearth extensions shall comply with UL 1618.

903.3 Unvented gas log heaters. An unvented gas log heater shall not be installed in a factory-built fireplace unless the fireplace system has been specifically tested, *listed* and *labeled* for such use in accordance with UL 127.

903.4 Gasketed fireplace doors. A gasketed fireplace door shall not be installed on a factory-built fireplace except where the fireplace system has been specifically tested, listed and labeled for such use in accordance with UL 127.

SECTION 904
PELLET FUEL-BURNING APPLIANCES

904.1 General. Pellet fuel-burning appliances shall be *listed* and *labeled* in accordance with ASTM E 1509 and shall be installed in accordance with the terms of the listing.

SECTION 905
FIREPLACE STOVES AND ROOM HEATERS

905.1 General. Fireplace stoves and solid-fuel-type room heaters shall be *listed* and *labeled* and shall be installed in accordance with the conditions of the listing. Fireplace stoves shall be tested in accordance with UL 737. Solid-fuel-type room heaters shall be tested in accordance with UL 1482. Fireplace inserts intended for installation in fireplaces shall be *listed* and *labeled* in accordance with the requirements of UL 1482 and shall be installed in accordance with the manufacturer's instructions.

905.2 Connection to fireplace. The connection of solid fuel appliances to *chimney* flues serving fireplaces shall comply with Sections 801.7 and 801.10.

905.3 Hearth extensions. Hearth extensions for fireplace stoves shall be installed in accordance with the listing of the fireplace stove. The hearth extension shall be readily distinguishable from the surrounding floor area. Listed and labeled hearth extensions shall comply with UL 1618.

SECTION 906
FACTORY-BUILT BARBECUE APPLIANCES

906.1 General. Factory-built barbecue appliances shall be of an *approved* type and shall be installed in accordance with the manufacturer's instructions, this chapter and Chapters 3, 5, 7 and 8, and the *International Fuel Gas Code*.

SECTION 907
INCINERATORS AND CREMATORIES

907.1 General. Incinerators and crematories shall be *listed* and *labeled* in accordance with UL 791 and shall be installed in accordance with the manufacturer's instructions.

SECTION 908
COOLING TOWERS, EVAPORATIVE CONDENSERS AND FLUID COOLERS

908.1 General. A cooling tower used in conjunction with an air-conditioning *appliance* shall be installed in accordance with the manufacturer's instructions. Factory-built cooling towers shall be listed in accordance with UL 1995.

908.2 Access. Cooling towers, evaporative condensers and fluid coolers shall be provided with ready access.

908.3 Location. Cooling towers, evaporative condensers and fluid coolers shall be located to prevent the discharge vapor plumes from entering occupied spaces. Plume discharges shall be not less than 5 feet (1524 mm) above or 20 feet (6096 mm) away from any ventilation inlet to a building. Location on the property shall be as required for buildings in accordance with the *International Building Code*.

908.4 Support and anchorage. Supports for cooling towers, evaporative condensers and fluid coolers shall be designed in accordance with the *International Building Code*. Seismic restraints shall be as required by the *International Building Code*.

908.5 Water supply. Cooling towers, evaporative coolers and fluid coolers shall be provided with an approved water supply, sized for peak demand. The quality of water shall be provided in accordance with the equipment manufacturer's recommendations. The piping system and protection of the potable water supply system shall be installed as required by the *International Plumbing Code*.

908.6 Drainage. Drains, overflows and blowdown provisions shall be indirectly connected to an *approved* disposal location. Discharge of chemical waste shall be *approved* by the appropriate regulatory authority.

908.7 Refrigerants and hazardous fluids. Heat exchange *equipment* that contains a refrigerant and that is part of a closed refrigeration system shall comply with Chapter 11. Heat exhange *equipment* containing heat transfer fluids which are flammable, combustible or hazardous shall comply with the *International Fire Code*.

908.8 Cooling towers. Cooling towers, both open circuit and closed circuit type, and evaporative condensers shall comply with Sections 908.8.1 and 908.8.2.

908.8.1 Conductivity or flow-based control of cycles of concentration. Cooling towers and evaporative condensers shall include controls that automate system bleed based on conductivity, fraction of metered makeup volume, metered bleed volume, recirculating pump run time or bleed time.

908.8.2 Drift eliminators. Cooling towers and evaporative condensers shall be equipped with drift eliminators that have a maximum drift rate of 0.005 percent of the circulated water flow rate as established in the equipment's design specifications.

SECTION 909
VENTED WALL FURNACES

909.1 General. Vented wall furnaces shall be installed in accordance with their listing and the manufacturer's instructions. Oil-fired furnaces shall be tested in accordance with UL 730.

909.2 Location. Vented wall furnaces shall be located so as not to cause a fire hazard to walls, floors, combustible furnishings or doors. Vented wall furnaces installed between bathrooms and adjoining rooms shall not circulate air from bathrooms to other parts of the building.

909.3 Door swing. Vented wall furnaces shall be located so that a door cannot swing within 12 inches (305 mm) of an air inlet or air outlet of such furnace measured at right angles to the opening. Doorstops or door closers shall not be installed to obtain this *clearance*.

909.4 Ducts prohibited. Ducts shall not be attached to wall furnaces. Casing extension boots shall not be installed unless *listed* as part of the *appliance*.

909.5 Manual shutoff valve. A manual shutoff valve shall be installed ahead of all controls.

909.6 Access. Vented wall furnaces shall be provided with access for cleaning of heating surfaces, removal of burners, replacement of sections, motors, controls, filters and other working parts, and for adjustments and lubrication of parts requiring such attention. Panels, grilles and access doors that must be removed for normal servicing operations shall not be attached to the building construction.

SECTION 910
FLOOR FURNACES

910.1 General. Floor furnaces shall be installed in accordance with their listing and the manufacturer's instructions. Oil-fired furnaces shall be tested in accordance with UL 729.

910.2 Placement. Floor furnaces shall not be installed in the floor of any aisle or passageway of any auditorium, public hall, place of assembly, or in any egress element from any such room or space.

With the exception of wall register models, a floor furnace shall not be placed closer than 6 inches (152 mm) to the nearest wall, and wall register models shall not be placed closer than 6 inches (152 mm) to a corner.

The furnace shall be placed such that a drapery or similar combustible object will not be nearer than 12 inches (305 mm) to any portion of the register of the furnace. Floor furnaces shall not be installed in concrete floor construction built on grade. The controlling thermostat for a floor furnace shall be located within the same room or space as the floor furnace or shall be located in an adjacent room or space that is permanently open to the room or space containing the floor furnace.

910.3 Bracing. The floor around the furnace shall be braced and headed with a support framework design in accordance with the *International Building Code*.

910.4 Clearance. The lowest portion of the floor furnace shall have not less than a 6-inch (152 mm) clearance from the grade level; except where the lower 6-inch (152 mm) portion of the floor furnace is sealed by the manufacturer to prevent entrance of water, the minimum clearance shall be reduced to not less than 2 inches (51 mm). Where these clearances are not present, the ground below and to the sides shall be excavated to form a pit under the furnace so that the required clearance is provided beneath the lowest portion of the furnace. A 12-inch (305 mm) minimum clearance shall be provided on all sides except the control side, which shall have an 18-inch (457 mm) minimum clearance.

SECTION 911
DUCT FURNACES

911.1 General. Duct furnaces shall be installed in accordance with the manufacturer's instructions. Electric duct furnaces shall comply with UL 1996.

SECTION 912
INFRARED RADIANT HEATERS

912.1 General. Electric infrared radiant heaters shall comply with UL 499.

912.2 Support. Infrared radiant heaters shall be fixed in a position independent of fuel and electric supply lines. Hangers and brackets shall be noncombustible material.

912.3 Clearances. Heaters shall be installed with clearances from combustible material in accordance with the manufacturer's installation instructions.

SECTION 913
CLOTHES DRYERS

913.1 General. Clothes dryers shall be installed in accordance with the manufacturer's instructions. Electric residential clothes dryers shall be tested in accordance with UL 2158. Electric coin-operated clothes dryers shall be tested in accordance with UL 2158. Electric commercial clothes dryers shall be tested in accordance with UL 1240.

913.2 Exhaust required. Clothes dryers shall be exhausted in accordance with Section 504.

913.3 Clearances. Clothes dryers shall be installed with *clearance* to combustibles in accordance with the manufacturer's instructions.

SECTION 914
SAUNA HEATERS

914.1 Location and protection. Sauna heaters shall be located so as to minimize the possibility of accidental contact by a person in the room.

914.1.1 Guards. Sauna heaters shall be protected from accidental contact by an *approved* guard or barrier of material having a low coefficient of thermal conductivity. The guard shall not substantially affect the transfer of heat from the heater to the room.

914.2 Installation. Sauna heaters shall be *listed* and *labeled* in accordance with UL 875 and shall be installed in accordance with their listing and the manufacturer's instructions.

914.3 Access. Panels, grilles and access doors that are required to be removed for normal servicing operations shall not be attached to the building.

914.4 Heat and time controls. Sauna heaters shall be equipped with a thermostat that will limit room temperature to 194ºF (90ºC). If the thermostat is not an integral part of the sauna heater, the heat-sensing element shall be located within 6 inches (152 mm) of the ceiling. If the heat-sensing element is a capillary tube and bulb, the assembly shall be attached to the wall or other support, and shall be protected against physical damage.

914.4.1 Timers. A timer, if provided to control main burner operation, shall have a maximum operating time of 1 hour. The control for the timer shall be located outside the sauna room.

914.5 Sauna room. A ventilation opening into the sauna room shall be provided. The opening shall be not less than 4 inches by 8 inches (102 mm by 203 mm) located near the top of the door into the sauna room.

914.5.1 Warning notice. The following permanent notice, constructed of *approved* material, shall be mechanically attached to the sauna room on the outside:

WARNING: DO NOT EXCEED 30 MINUTES IN SAUNA. EXCESSIVE EXPOSURE CAN BE HARMFUL TO HEALTH. ANY PERSON WITH POOR HEALTH SHOULD CONSULT A PHYSICIAN BEFORE USING SAUNA.

The words shall contrast with the background and the wording shall be in letters not less than $^1/_4$-inch (6.4 mm) high.

Exception: This section shall not apply to one- and two-family dwellings.

SECTION 915
ENGINE AND GAS TURBINE-POWERED
EQUIPMENT AND APPLIANCES

915.1 General. The installation of liquid-fueled stationary internal *combustion* engines and gas turbines, including exhaust, fuel storage and piping, shall meet the requirements of NFPA 37. Stationary engine generator assemblies shall meet the requirements of UL 2200.

915.2 Powered equipment and appliances. Permanently installed *equipment* and appliances powered by internal *combustion* engines and turbines shall be installed in accordance with the manufacturer's instructions and NFPA 37.

SECTION 916
POOL AND SPA HEATERS

916.1 General. Pool and spa heaters shall be installed in accordance with the manufacturer's instructions. Oil-fired pool and spa heaters shall be tested in accordance with UL

726. Electric pool and spa heaters shall be tested in accordance with UL 1261.

SECTION 917
COOKING APPLIANCES

917.1 Cooking appliances. Cooking appliances that are designed for permanent installation, including ranges, ovens, stoves, broilers, grills, fryers, griddles and barbecues, shall be *listed*, *labeled* and installed in accordance with the manufacturer's instructions. Commercial electric cooking appliances shall be *listed* and *labeled* in accordance with UL 197. Household electric ranges shall be *listed* and *labeled* in accordance with UL 858. Microwave cooking appliances shall be *listed* and *labeled* in accordance with UL 923. Oil-burning stoves shall be *listed* and *labeled* in accordance with UL 896. Solid-fuel-fired ovens shall be *listed* and *labeled* in accordance with UL 2162.

917.2 Domestic appliances. Cooking appliances installed within *dwelling units* and within areas where domestic cooking operations occur shall be *listed* and *labeled* as household-type appliances for domestic use.

SECTION 918
FORCED-AIR WARM-AIR FURNACES

918.1 Forced-air furnaces. Oil-fired furnaces shall be tested in accordance with UL 727. Electric furnaces shall be tested in accordance with UL 1995. Solid fuel furnaces shall be tested in accordance with UL 391. Forced-air furnaces shall be installed in accordance with the listings and the manufacturer's instructions.

918.2 Heat pumps. Electric heat pumps shall be tested in accordance with UL 1995.

918.3 Dampers. Volume dampers shall not be placed in the air inlet to a furnace in a manner that will reduce the required air to the furnace.

918.4 Circulating air ducts for forced-air warm-air furnaces. Circulating air for fuel-burning, forced-air-type, warm-air furnaces shall be conducted into the blower housing from outside the furnace enclosure by continuous air-tight ducts.

918.5 Outdoor and return air openings. Outdoor intake openings shall be located in accordance with Section 401.4. Return air openings shall be located in accordance with Section 601.5.

918.6 Outdoor opening protection. Outdoor air intake openings shall be protected in accordance with Section 401.5.

SECTION 919
CONVERSION BURNERS

919.1 Conversion burners. The installation of conversion burners shall conform to ANSI Z21.8.

SECTION 920
UNIT HEATERS

920.1 General. Unit heaters shall be installed in accordance with the listing and the manufacturer's instructions. Oil-fired unit heaters shall be tested in accordance with UL 731.

920.2 Support. Suspended-type unit heaters shall be supported by elements that are designed and constructed to accommodate the weight and dynamic loads. Hangers and brackets shall be of noncombustible material. Suspended-type oil-fired unit heaters shall be installed in accordance with NFPA 31.

920.3 Ductwork. A unit heater shall not be attached to a warm-air duct system unless *listed* for such installation.

SECTION 921
VENTED ROOM HEATERS

921.1 General. Vented room heaters shall be *listed* and *labeled* and shall be installed in accordance with the conditions of the listing and the manufacturer's instructions.

SECTION 922
KEROSENE AND OIL-FIRED STOVES

922.1 General. Kerosene and oil-fired stoves shall be listed and labeled and shall be installed in accordance with the conditions of the listing and the manufacturer's instructions. Kerosene and oil-fired stoves shall comply with NFPA 31 and UL 896.

SECTION 923
SMALL CERAMIC KILNS

923.1 General. Kilns shall be listed and labeled unless otherwise approved in accordance with Section 105.2. Electric kilns shall comply with UL 499. The approval of unlisted appliances in accordance with Section 105.2 shall be based on approved engineering evaluation.

923.1.1 Installation. Kilns shall be installed in accordance with the manufacturer's instructions and the provisions of this code.

SECTION 924
STATIONARY FUEL CELL POWER SYSTEMS

924.1 General. Stationary fuel cell power systems having a power output not exceeding 10 MW shall be tested in accordance with ANSI/CSA America FC 1 and shall be installed in accordance with the manufacturer's instructions, NFPA 853, the *International Building Code* and the *International Fire Code*.

SECTION 925
MASONRY HEATERS

925.1 General. Masonry heaters shall be constructed in accordance with the *International Building Code*.

SECTION 926
GASEOUS HYDROGEN SYSTEMS

926.1 Installation. The installation of gaseous hydrogen systems shall be in accordance with the applicable requirements of this code, the *International Fire Code*, the *International Fuel Gas Code* and the *International Building Code*.

SECTION 927
RADIANT HEATING SYSTEMS

927.1 General. Electric radiant heating systems shall be installed in accordance with the manufacturer's instructions and shall be listed for the application.

927.2 Clearances. Clearances for radiant heating panels or elements to any wiring, outlet boxes and junction boxes used for installing electrical devices or mounting luminaires shall be in accordance with the *International Building Code* and NFPA 70.

927.3 Installation on wood or steel framing. Radiant panels installed on wood or steel framing shall conform to the following requirements:

1. Heating panels shall be installed parallel to framing members and secured to the surface of framing members or shall be mounted between framing members.

2. Mechanical fasteners shall penetrate only the unheated portions provided for this purpose. Panels shall not be fastened at any point closer than $^1/_4$ inch (6.4 mm) to an element. Other methods of attachment of the panels shall be in accordance with the panel installation instructions.

3. Unless listed and labeled for field cutting, heating panels shall be installed as complete units.

927.4 Installation in concrete or masonry. Radiant heating systems installed in concrete or masonry shall conform to the following requirements:

1. Radiant heating systems shall be identified as being suitable for the installation, and shall be secured in place as specified in the manufacturer's instructions.

2. Radiant heating panels and radiant heating panel sets shall not be installed where they bridge expansion joints unless they are protected from expansion and contraction.

927.5 Finish surfaces. Finish materials installed over radiant heating panels and systems shall be installed in accordance with the manufacturer's instructions. Surfaces shall be secured so that fasteners do not pierce the radiant heating elements.

SECTION 928
EVAPORATIVE COOLING EQUIPMENT

928.1 General. Evaporative cooling equipment shall:

1. Be installed in accordance with the manufacturer's instructions.

2. Be installed on level platforms in accordance with Section 304.10.

3. Have openings in exterior walls or roofs flashed in accordance with the *International Building Code*.

4. Be provided with an approved water supply, sized for peak demand. The quality of water shall be provided in accordance with the equipment manufacturer's recommendations. The piping system and protection of the potable water supply system shall be installed as required by the *International Plumbing Code*.

5. Have air intake opening locations in accordance with Section 401.4.

CHAPTER 10

BOILERS, WATER HEATERS AND PRESSURE VESSELS

SECTION 1001
GENERAL

1001.1 Scope. This chapter shall govern the installation, *alteration* and repair of boilers, water heaters and pressure vessels.

Exceptions:

1. Pressure vessels used for unheated water supply.

2. Portable unfired pressure vessels and Interstate Commerce Commission containers.

3. Containers for bulk oxygen and medical gas.

4. Unfired pressure vessels having a volume of 5 cubic feet (0.14 m^3) or less operating at pressures not exceeding 250 pounds per square inch (psi) (1724 kPa) and located within occupancies of Groups B, F, H, M, R, S and U.

5. Pressure vessels used in refrigeration systems that are regulated by Chapter 11 of this code.

6. Pressure tanks used in conjunction with coaxial cables, telephone cables, power cables and other similar humidity control systems.

7. Any boiler or pressure vessel subject to inspection by federal or state inspectors.

SECTION 1002
WATER HEATERS

1002.1 General. Potable water heaters and hot water storage tanks shall be listed and labeled and installed in accordance with the manufacturer's instructions, the *International Plumbing Code* and this code. All water heaters shall be capable of being removed without first removing a permanent portion of the building structure. The potable water connections and relief valves for all water heaters shall conform to the requirements of the *International Plumbing Code*. Domestic electric water heaters shall comply with UL 174 or UL 1453. Commercial electric water heaters shall comply with UL 1453. Oil-fired water heaters shall comply with UL 732. Solid-fuel-fired water heaters shall comply with UL 2523. Thermal solar water heaters shall comply with Chapter 14 and UL 174 or UL 1453.

1002.2 Water heaters utilized for space heating. Water heaters utilized both to supply potable hot water and provide hot water for space-heating applications shall be *listed* and *labeled* for such applications by the manufacturer and shall be installed in accordance with the manufacturer's instructions and the *International Plumbing Code*.

1002.2.1 Sizing. Water heaters utilized for both potable water heating and space-heating applications shall be sized to prevent the space-heating load from diminishing the required potable water-heating capacity.

1002.2.2 Temperature limitation. Where a combination potable water-heating and space-heating system requires water for space heating at temperatures higher than 140°F (60°C), a temperature-actuated mixing valve that conforms to ASSE 1017 shall be provided to temper the water supplied to the potable hot water distribution system to a temperature of 140°F (60°C) or less.

1002.3 Supplemental water-heating devices. Potable water-heating devices that utilize refrigerant-to-water heat exchangers shall be *approved* and installed in accordance with the *International Plumbing Code* and the manufacturer's instructions.

SECTION 1003
PRESSURE VESSELS

1003.1 General. All pressure vessels, unless otherwise approved, shall be constructed and certified in accordance with the ASME *Boiler and Pressure Vessel Code*, and shall be installed in accordance with the manufacturer's instructions and nationally recognized standards. Directly fired pressure vessels shall meet the requirements of Section 1004.

1003.2 Piping. All piping materials, fittings, joints, connections and devices associated with systems utilized in conjunction with pressure vessels shall be designed for the specific application and shall be *approved*.

1003.3 Welding. Welding on pressure vessels shall be performed by an R-Stamp holder in accordance with the *National Board Inspection Code, Part 3* or in accordance with an *approved* standard.

SECTION 1004
BOILERS

1004.1 Standards. Boilers shall be designed, constructed and certified in accordance with the ASME *Boiler and Pressure Vessel Code*, Section I or IV. Controls and safety devices for boilers with fuel input ratings of 12,500,000 Btu/hr (3,662,500 W) or less shall meet the requirements of ASME CSD-1. Controls and safety devices for boilers with inputs greater than 12,500,000 Btu/hr (3,662,500 W) shall meet the requirements of NFPA 85. Packaged oil-fired boilers shall be listed and labeled in accordance with UL 726. Packaged electric boilers shall be listed and labeled in accordance with UL 834. Solid-fuel-fired boilers shall be listed and labeled in accordance with UL 2523.

1004.2 Installation. In addition to the requirements of this code, the installation of boilers shall conform to the manufacturer's instructions. Operating instructions of a permanent type shall be attached to the boiler. Boilers shall have all controls set, adjusted and tested by the installer. The manufacturer's rating data and the nameplate shall be attached to the boiler.

1004.3 Working clearance. Clearances shall be maintained around boilers, generators, heaters, tanks and related *equipment* and appliances so as to permit inspection, servicing, repair, replacement and visibility of all gauges. When boilers are installed or replaced, clearance shall be provided to allow access for inspection, maintenance and repair. Passageways around all sides of boilers shall have an unobstructed width of not less than 18 inches (457 mm), unless otherwise *approved*.

1004.3.1 Top clearance. Clearances from the tops of boilers to the ceiling or other overhead obstruction shall be in accordance with Table 1004.3.1.

TABLE 1004.3.1
BOILER TOP CLEARANCES

BOILER TYPE	MINIMUM CLEARANCES FROM TOP OF BOILER TO CEILING OR OTHER OVERHEAD OBSTRUCTION (feet)
All boilers with manholes on top of the boiler except where a greater clearance is required in this table.	3
All boilers without manholes on top of the boiler except high-pressure steam boilers and where a greater clearance is required in this table.	2
High-pressure steam boilers with steam generating capacity not exceeding 5,000 pounds per hour.	3
High-pressure steam boilers with steam generating capacity exceeding 5,000 pounds per hour.	7
High-pressure steam boilers having heating surface not exceeding 1,000 square feet.	3
High-pressure steam boilers having heating surface in excess of 1,000 square feet.	7
High-pressure steam boilers with input not exceeding 5,000,000 Btu/h.	3
High-pressure steam boilers with input in excess of 5,000,000 Btu/h.	7
Steam-heating boilers and hot water-heating boilers with input exceeding 5,000,000 Btu/h.	3
Steam-heating boilers exceeding 5,000 pounds of steam per hour.	3
Steam-heating boilers and hot water-heating boilers having heating surface exceeding 1,000 square feet.	3

For SI: 1 foot = 304.8 mm, 1 square foot = 0.0929 m², 1 pound per hour = 0.4536 kg/h, 1 Btu/hr = 0.293 W.

1004.4 Mounting. *Equipment* shall be set or mounted on a level base capable of supporting and distributing the weight contained thereon. Boilers, tanks and *equipment* shall be secured in accordance with the manufacturer's installation instructions.

1004.5 Floors. Boilers shall be mounted on floors of noncombustible construction, unless *listed* for mounting on combustible flooring.

1004.6 Boiler rooms and enclosures. Boiler rooms and enclosures and access thereto shall comply with the *International Building Code* and Chapter 3 of this code. Boiler rooms shall be equipped with a floor drain or other *approved* means for disposing of liquid waste.

1004.7 Operating adjustments and instructions. Hot water and steam boilers shall have all operating and safety controls set and operationally tested by the installing contractor. A complete control diagram and boiler operating instructions shall be furnished by the installer for each installation.

SECTION 1005
BOILER CONNECTIONS

1005.1 Valves. Every boiler or modular boiler shall have a shutoff valve in the supply and return piping. For multiple boiler or multiple modular boiler installations, each boiler or modular boiler shall have individual shutoff valves in the supply and return piping.

Exception: Shutoff valves are not required in a system having a single low-pressure steam boiler.

1005.2 Potable water supply. The water supply to all boilers shall be connected in accordance with the *International Plumbing Code*.

SECTION 1006
SAFETY AND PRESSURE RELIEF
VALVES AND CONTROLS

1006.1 Safety valves for steam boilers. Steam boilers shall be protected with a safety valve.

1006.2 Safety relief valves for hot water boilers. Hot water boilers shall be protected with a safety relief valve.

1006.3 Pressure relief for pressure vessels. Pressure vessels shall be protected with a pressure relief valve or pressure-limiting device as required by the manufacturer's installation instructions for the pressure vessel.

1006.4 Approval of safety and safety relief valves. Safety and safety relief valves shall be *listed* and *labeled*, and shall have a minimum rated capacity for the *equipment* or appliances served. Safety and safety relief valves shall be set at not greater than the nameplate pressure rating of the boiler or pressure vessel.

1006.5 Installation. Safety or relief valves shall be installed directly into the safety or relief valve opening on the boiler or pressure vessel. Valves shall not be located on either side of a safety or relief valve connection. The relief valve shall discharge by gravity.

1006.6 Safety and relief valve discharge. Safety and relief valve discharge pipes shall be of rigid pipe that is *approved* for the temperature of the system. The discharge pipe shall be the same diameter as the safety or relief valve outlet. Safety and relief valves shall not discharge so as to be a hazard, a potential cause of damage or otherwise a nuisance. High-pressure-steam safety valves shall be vented to the outside of the structure. Where a low-pressure safety valve or a relief valve discharges to the drainage system, the installation shall conform to the *International Plumbing Code*.

1006.7 Boiler safety devices. Boilers shall be equipped with controls and limit devices as required by the manufacturer's installation instructions and the conditions of the listing.

1006.8 Electrical requirements. The power supply to the electrical control system shall be from a two-wire branch circuit that has a grounded conductor, or from an isolation transformer with a two-wire secondary. Where an isolation transformer is provided, one conductor of the secondary winding shall be grounded. Control voltage shall not exceed 150 volts nominal, line to line. Control and limit devices shall interrupt the ungrounded side of the circuit. A means of manually disconnecting the control circuit shall be provided and controls shall be arranged so that when deenergized, the burner shall be inoperative. Such disconnecting means shall be capable of being locked in the off position and shall be provided with ready access.

SECTION 1007
BOILER LOW-WATER CUTOFF

1007.1 General. Steam and hot water boilers shall be protected with a low-water cutoff control.

> **Exception:** A low-water cutoff is not required for coil-type and water-tube-type boilers that require forced circulation of water through the boiler and that are protected with a flow sensing control.

1007.2 Operation. Low-water cutoff controls and flow sensing controls required by Section 1007.1 shall automatically stop the *combustion* operation of the *appliance* when the water level drops below the lowest safe water level as established by the manufacturer or when water circulation stops, respectively.

SECTION 1008
BOTTOM BLOWOFF VALVE

1008.1 General. Steam boilers shall be equipped with bottom blowoff valve(s). The valve(s) shall be installed in the opening provided on the boiler. The minimum size of the valve(s) and associated piping shall be the size specified by the boiler manufacturer or the size of the boiler blowoff-valve opening. Where the maximum allowable working pressure of the boiler exceeds 100 psig (689 kPa), two bottom blowoff valves shall be provided consisting of either two slow-opening valves in series or one quick-opening valve and one slow-opening valve in series, with the quick-opening valve installed closest to the boiler.

1008.2 Discharge. Blowoff valves shall discharge to a safe place of disposal. Where discharging to the drainage system, the installation shall conform to the *International Plumbing Code*.

SECTION 1009
HOT WATER BOILER EXPANSION TANK

1009.1 Where required. An expansion tank shall be installed in every hot water system. For multiple boiler installations, not less than one expansion tank is required. Expansion tanks shall be of the closed or open type. Tanks shall be rated for the pressure of the hot water system.

1009.2 Closed-type expansion tanks. Closed-type expansion tanks shall be installed in accordance with the manufacturer's instructions. Expansion tanks for systems designed to have an operating pressure in excess of 30 psi (207 kPa) shall be constructed and certified in accordance with the ASME *Boiler and Pressure Vessel Code*. The size of the tank shall be based on the capacity of the hot-water-heating system. The minimum size of the tank shall be determined in accordance with the following equation where all necessary information is known:

$$V_t = \frac{(0.00041\,T - 0.0466)V_s}{\left(\frac{P_a}{P_f}\right) - \left(\frac{P_a}{P_o}\right)}$$

(Equation 10-1)

For SI:

$$V_t = \frac{(0.000738\,T - 0.03348)V_s}{\left(\frac{P_a}{P_f}\right) - \left(\frac{P_a}{P_o}\right)}$$

where:

V_t = Minimum volume of tanks (gallons) (L).

V_s = Volume of system, not including expansion tanks (gallons) (L).

T = Average operating temperature (°F) (°C).

P_a = Atmospheric pressure (psi) (kPa).

P_f = Fill pressure (psi) (kPa).

P_o = Maximum operating pressure (psi) (kPa).

Where all necessary information is not known, the minimum size of the tank shall be determined from Table 1009.2.

TABLE 1009.2
CLOSED-TYPE EXPANSION TANK SIZING

SYSTEM VOLUME IN GALLONS	TANK CAPACITIES IN GALLONS	
	Pressurized Diaphragm Type	Nonpressurized Type
100	9	15
200	17	30
300	25	45
400	33	60
500	42	75
1,000	83	150
2,000	165	300

For SI: 1 gallon = 3.795 L.

1009.3 Open-type expansion tanks. Open-type expansion tanks shall be located not less than 4 feet (1219 mm) above the highest heating element. The tank shall be adequately sized for the hot water system. An overflow with a minimum diameter of 1 inch (25 mm) shall be installed at the top of the tank. The overflow shall discharge to the drainage system in accordance with the *International Plumbing Code*.

SECTION 1010
GAUGES

1010.1 Hot water boiler gauges. Every hot water boiler shall have a pressure gauge and a temperature gauge, or a combination pressure and temperature gauge. The gauges shall indicate the temperature and pressure within the normal range of the system's operation.

1010.2 Steam boiler gauges. Every steam boiler shall have a water-gauge glass and a pressure gauge. The pressure gauge shall indicate the pressure within the normal range of the system's operation.

1010.2.1 Water-gauge glass. The gauge glass shall be installed so that the midpoint is at the normal boiler water level.

SECTION 1011
TESTS

1011.1 Tests. Upon completion of the assembly and installation of boilers and pressure vessels, acceptance tests shall be conducted in accordance with the requirements of the ASME *Boiler and Pressure Vessel Code* or the manufacturer's requirements, and such tests shall be approved. A copy of all test documents along with all manufacturer's data reports required by the ASME *Boiler and Pressure Vessel Code* shall be submitted to the code official.

1011.2 Test gauges. An indicating test gauge shall be connected directly to the boiler or pressure vessel where it is visible to the operator throughout the duration of the test. The pressure gauge scale shall be graduated over a range of not less than one and one-half times and not greater than four times the maximum test pressure. Gauges utilized for testing shall be calibrated and certified by the test operator.

CHAPTER 11

REFRIGERATION

SECTION 1101
GENERAL

1101.1 Scope. This chapter shall govern the design, installation, construction and repair of refrigeration systems that vaporize and liquefy a fluid during the refrigerating cycle. Refrigerant piping design and installation, including pressure vessels and pressure relief devices, shall conform to this code. Permanently installed refrigerant storage systems and other components shall be considered as part of the refrigeration system to which they are attached.

1101.2 Factory-built equipment and appliances. *Listed* and *labeled* self-contained, factory-built *equipment* and appliances shall be tested in accordance with UL 207, 412, 471 or 1995. Such *equipment* and appliances are deemed to meet the design, manufacture and factory test requirements of this code if installed in accordance with their listing and the manufacturer's instructions.

1101.3 Protection. Any portion of a refrigeration system that is subject to physical damage shall be protected in an *approved* manner.

1101.4 Water connection. Water supply and discharge connections associated with refrigeration systems shall be made in accordance with this code and the *International Plumbing Code*.

1101.5 Fuel gas connection. Fuel gas devices, *equipment* and appliances used with refrigeration systems shall be installed in accordance with the *International Fuel Gas Code*.

1101.6 General. Refrigeration systems shall comply with the requirements of this code and, except as modified by this code, ASHRAE 15. Ammonia-refrigerating systems shall comply with this code and, except as modified by this code, ASHRAE 15 and IIAR 2.

1101.7 Maintenance. Mechanical refrigeration systems shall be maintained in proper operating condition, free from accumulations of oil, dirt, waste, excessive corrosion, other debris and leaks.

1101.8 Change in refrigerant type. The type of refrigerant in refrigeration systems having a refrigerant circuit containing more than 220 pounds (99.8 kg) of Group A1 or 30 pounds (13.6 kg) of any other group refrigerant shall not be changed without prior notification to the code official and compliance with the applicable code provisions for the new refrigerant type.

[F] 1101.9 Refrigerant discharge. Notification of refrigerant discharge shall be provided in accordance with the *International Fire Code*.

1101.10 Locking access port caps. Refrigerant circuit access ports located outdoors shall be fitted with locking-type tamper-resistant caps or shall be otherwise secured to prevent unauthorized access.

Exception: This section shall not apply to refrigerant circuit access ports on equipment installed in controlled areas such as on roofs with locked access hatches or doors.

SECTION 1102
SYSTEM REQUIREMENTS

1102.1 General. The system classification, allowable refrigerants, maximum quantity, enclosure requirements, location limitations, and field pressure test requirements shall be determined as follows:

1. Determine the refrigeration system's classification, in accordance with Section 1103.3.

2. Determine the refrigerant classification in accordance with Section 1103.1.

3. Determine the maximum allowable quantity of refrigerant in accordance with Section 1104, based on type of refrigerant, system classification and *occupancy*.

4. Determine the system enclosure requirements in accordance with Section 1104.

5. Refrigeration *equipment* and *appliance* location and installation shall be subject to the limitations of Chapter 3.

6. Nonfactory-tested, field-erected *equipment* and appliances shall be pressure tested in accordance with Section 1108.

1102.2 Refrigerants. The refrigerant shall be that which the *equipment* or *appliance* was designed to utilize or converted to utilize. Refrigerants not identified in Table 1103.1 shall be *approved* before use.

1102.2.1 Mixing. Refrigerants, including refrigerant blends, with different designations in ASHRAE 34 shall not be mixed in a system.

Exception: Addition of a second refrigerant is allowed where permitted by the *equipment* or *appliance* manufacturer to improve oil return at low temperatures. The refrigerant and amount added shall be in accordance with the manufacturer's instructions.

1102.2.2 Purity. Refrigerants used in refrigeration systems shall be new, recovered or *reclaimed refrigerants* in accordance with Section 1102.2.2.1, 1102.2.2.2 or

1102.2.2.3. Where required by the *equipment* or *appliance* owner or the code official, the installer shall furnish a signed declaration that the refrigerant used meets the requirements of Section 1102.2.2.1, 1102.2.2.2 or 1102.2.2.3.

> **Exception:** The refrigerant used shall meet the purity specifications set by the manufacturer of the *equipment* or *appliance* in which such refrigerant is used where such specifications are different from that specified in Sections 1102.2.2.1, 1102.2.2.2 and 1102.2.2.3.

1102.2.2.1 New refrigerants. Refrigerants shall be of a purity level specified by the *equipment* or *appliance* manufacturer.

1102.2.2.2 Recovered refrigerants. Refrigerants that are recovered from refrigeration and air-conditioning systems shall not be reused in other than the system from which they were recovered and in other systems of the same owner. *Recovered refrigerants* shall be filtered and dried before reuse. *Recovered refrigerants* that show clear signs of contamination shall not be reused unless reclaimed in accordance with Section 1102.2.2.3.

1102.2.2.3 Reclaimed refrigerants. Used refrigerants shall not be reused in a different owner's *equipment* or appliances unless tested and found to meet the purity requirements of ARI 700. Contaminated refrigerants shall not be used unless reclaimed and found to meet the purity requirements of ARI 700.

1102.3 Access port protection. Refrigerant access ports shall be protected in accordance with Section 1101.10 whenever refrigerant is added to or recovered from refrigeration or air-conditioning systems.

SECTION 1103
REFRIGERATION SYSTEM CLASSIFICATION

1103.1 Refrigerant classification. Refrigerants shall be classified in accordance with ASHRAE 34 as listed in Table 1103.1.

1103.2 Occupancy classification. Locations of refrigerating systems are described by *occupancy* classifications that consider the ability of people to respond to potential exposure to refrigerants. Where *equipment* or appliances, other than piping, are located outside a building and within 20 feet (6096 mm) of any building opening, such *equipment* or appliances shall be governed by the *occupancy* classification of the building. *Occupancy* classifications shall be defined as follows:

1. Institutional *occupancy* is that portion of premises from which occupants cannot readily leave without the assistance of others because they are disabled, debilitated or confined. Institutional occupancies include, among others, hospitals, nursing homes, asylums and spaces containing locked cells.

2. Public assembly *occupancy* is that portion of premises where large numbers of people congregate and from which occupants cannot quickly vacate the space. Public assembly occupancies include, among others, auditoriums, ballrooms, classrooms, passenger depots, restaurants and theaters.

3. Residential *occupancy* is that portion of premises that provides the occupants with complete independent living facilities, including permanent provisions for living, sleeping, eating, cooking and sanitation. Residential occupancies include, among others, dormitories, hotels, multiunit apartments and private residences.

4. Commercial *occupancy* is that portion of premises where people transact business, receive personal service or purchase food and other goods. Commercial occupancies include, among others, office and professional buildings, markets (but not large mercantile occupancies) and work or storage areas that do not qualify as industrial occupancies.

5. Large mercantile *occupancy* is that portion of premises where more than 100 persons congregate on levels above or below street level to purchase personal merchandise.

6. Industrial *occupancy* is that portion of premises that is not open to the public, where access by authorized persons is controlled, and that is used to manufacture, process or store goods such as chemicals, food, ice, meat or petroleum.

7. Mixed *occupancy* occurs where two or more occupancies are located within the same building. Where each *occupancy* is isolated from the rest of the building by tight walls, floors and ceilings and by self-closing doors, the requirements for each *occupancy* shall apply to its portion of the building. Where the various occupancies are not so isolated, the *occupancy* having the most stringent requirements shall be the governing *occupancy*.

1103.3 System classification. Refrigeration systems shall be classified according to the degree of probability that refrigerant leaked from a failed connection, seal or component could enter an occupied area. The distinction is based on the basic design or location of the components.

1103.3.1 Low-probability systems. Double-indirect open-spray systems, indirect closed systems and indirect-vented closed systems shall be classified as low-probability systems, provided that all refrigerant-containing piping and fittings are isolated when the quantities in Table 1103.1 are exceeded.

1103.3.2 High-probability systems. Direct systems and indirect open-spray systems shall be classified as high-probability systems.

> **Exception:** An indirect open-spray system shall not be required to be classified as a high-probability system if the pressure of the secondary coolant is at all times (operating and standby) greater than the pressure of the refrigerant.

TABLE 1103.1
REFRIGERANT CLASSIFICATION, AMOUNT AND OEL

CHEMICAL REFRIGERANT	FORMULA	CHEMICAL NAME OF BLEND	REFRIGERANT CLASSIFICATION	AMOUNT OF REFRIGERANT PER OCCUPIED SPACE				[F] DEGREES OF HAZARD[a]
				Pounds per 1,000 cubic feet	ppm	g/m³	OEL[e]	
R-11[d]	CCl_3F	trichlorofluoromethane	A1	0.39	1,100	6.2	C1,000	2-0-0[b]
R-12[d]	CCl_2F_2	dichlorodifluoromethane	A1	5.6	18,000	90	1,000	2-0-0[b]
R-13[d]	$CClF_3$	chlorotrifluoromethane	A1	—	—	—	1,000	2-0-0[b]
R-13B1[d]	$CBrF_3$	bromotrifluoromethane	A1	—	—	—	1,000	2-0-0[b]
R-14	CF_4	tetrafluoromethane (carbon tetrafluoride)	A1	25	110,000	400	1,000	2-0-0[b]
R-22	$CHClF_2$	chlorodifluoromethane	A1	13	59,000	210	1,000	2-0-0[b]
R-23	CHF_3	trifluoromethane (fluoroform)	A1	7.3	41,000	120	1,000	2-0-0[b]
R-32	CH_2F_2	difluoromethane (methylene fluoride)	A2[f]	4.8	36,000	77	1,000	1-4-0
R-113[d]	CCl_2FCClF_2	1,1,2-trichloro-1,2,2-trifluoroethane	A1	1.2	2,600	20	1,000	2-0-0[b]
R-114[d]	$CClF_2CClF_2$	1,2-dichloro-1,1,2,2-tetrafluoroethane	A1	8.7	20,000	140	1,000	2-0-0[b]
R-115	$CClF_2CF_3$	chloropentafluoroethane	A1	47	120,000	760	1,000	—
R-116	CF_3CF_3	hexafluoroethane	A1	34	97,000	550	1,000	1-0-0
R-123	$CHCl_2CF_3$	2,2-dichloro-1,1,1-trifluoroethane	B1	3.5	9,100	57	50	2-0-0[b]
R-124	$CHClFCF_3$	2-chloro-1,1,1,2-tetrafluoroethane	A1	3.5	10,000	56	1,000	2-0-0[b]
R-125	CHF_2CF_3	pentafluoroethane	A1	23	75,000	370	1,000	2-0-0[b]
R-134a	CH_2FCF_3	1,1,1,2-tetrafluoroethane	A1	13	50,000	210	1,000	2-0-0[b]
R-141b	CH_3CCl_2F	1,1-dichloro-1-fluoroethane	—	0.78	2,600	12	500	2-1-0
R-142b	CH_3CClF_2	1-chloro-1,1-difluoroethane	A2	5.1	20,000	83	1,000	2-4-0
R-143a	CH_3CF_3	1,1,1-trifluoroethane	A2[f]	4.5	21,000	70	1,000	2-0-0[b]
R-152a	CH_3CHF_2	1,1-difluoroethane	A2	2.0	12,000	32	1,000	1-4-0
R-170	CH_3CH_3	ethane	A3	0.54	7,000	8.7	1,000	2-4-0
R-E170	CH_3OCH_3	Methoxymethane (dimethyl ether)	A3	1.0	8,500	16	1,000	—
R-218	$CF_3CF_2CF_3$	octafluoropropane	A1	43	90,000	690	1,000	2-0-0[b]
R-227ea	CF_3CHFCF_3	1,1,1,2,3,3,3-heptafluoropropane	A1	36	84,000	580	1,000	—
R-236fa	$CF_3CH_2CF_3$	1,1,1,3,3,3-hexafluoropropane	A1	21	55,000	340	1,000	2-0-0[b]
R-245fa	$CHF_2CH_2CF_3$	1,1,1,3,3-pentafluoropropane	B1	12	34,000	190	300	2-0-0[b]
R-290	$CH_3CH_2CH_3$	propane	A3	0.56	5,300	9.5	1,000	2-4-0
R-C318	$-(CF_2)_4-$	octafluorocyclobutane	A1	41	80,000	660	1,000	—
R-400[d]	zeotrope	R-12/114 (50.0/50.0)	A1	10	28,000	160	1,000	2-0-0[b]
R-400[d]	zeotrope	R-12/114 (60.0/40.0)	A1	11	30,000	170	1,000	
R-401A	zeotrope	R-22/152a/124 (53.0/13.0/34.0)	A1	6.6	27,000	110	1,000	2-0-0[b]

(continued)

TABLE 1103.1—continued
REFRIGERANT CLASSIFICATION, AMOUNT AND OEL

CHEMICAL REFRIGERANT	FORMULA	CHEMICAL NAME OF BLEND	REFRIGERANT CLASSIFICATION	AMOUNT OF REFRIGERANT PER OCCUPIED SPACE				[F] DEGREES OF HAZARD[a]
				Pounds per 1,000 cubic feet	ppm	g/m³	OEL[e]	
R-401B	zeotrope	R-22/152a/124 (61.0/11.0/28.0)	A1	7.2	30,000	120	1,000	2-0-0[b]
R-401C	zeotrope	R-22/152a/124 (33.0/15.0/52.0)	A1	5.2	20,000	84	1,000	2-0-0[b]
R-402A	zeotrope	R-125/290/22 (60.0/2.0/38.0)	A1	17	66,000	270	1,000	2-0-0[b]
R-402B	zeotrope	R-125/290/22 (38.0/2.0/60.0)	A1	15	63,000	240	1,000	2-0-0[b]
R-403A	zeotrope	R-290/22/218 (5.0/75.0/20.0)	A2	7.6	33,000	120	1,000	2-0-0[b]
R-403B	zeotrope	R-290/22/218 (5.0/56.0/39.0)	A1	18	70,000	290	1,000	2-0-0[b]
R-404A	zeotrope	R-125/143a/134a (44.0/52.0/4.0)	A1	31	130,000	500	1,000	2-0-0[b]
R-405A	zeotrope	R-22/152a/142b/C318 (45.0/7.0/5.5/2.5)	—	16	57,000	260	1,000	—
R-406A	zeotrope	R-22/600a/142b (55.0/4.0/41.0)	A2	4.7	21,000	25	1,000	—
R-407A	zeotrope	R-32/125/134a (20.0/40.0/40.0)	A1	19	83,000	300	1,000	2-0-0[b]
R-407B	zeotrope	R-32/125/134a (10.0/70.0/20.0)	A1	21	79,000	330	1,000	2-0-0[b]
R-407C	zeotrope	R-32/125/134a (23.0/25.0/52.0)	A1	18	81,000	290	1,000	2-0-0[b]
R-407D	zeotrope	R-32/125/134a (15.0/15.0/70.0)	A1	16	68,000	250	1,000	2-0-0[b]
R-407E	zeotrope	R-32/125/134a (25.0/15.0/60.0)	A1	17	80,000	280	1,000	2-0-0[b]
R-407F	zeotrope	R-32/125/134a (30.0/30.0/40.0)	A1	20	95,000	320	1,000	—
R-408A	zeotrope	R-125/143a/22 (7.0/46.0/47.0)	A1	21	95,000	340	1,000	2-0-0[b]
R-409A	zeotrope	R-22/124/142b (60.0/25.0/15.0)	A1	7.1	29,000	110	1,000	2-0-0[b]
R-409B	zeotrope	R-22/124/142b (65.0/25.0/10.0)	A1	7.3	30,000	120	1,000	2-0-0[b]
R-410A	zeotrope	R-32/125 (50.0/50.0)	A1	26	140,000	420	1,000	2-0-0[b]
R-410B	zeotrope	R-32/125 (45.0/55.0)	A1	27	140,000	430	1,000	2-0-0[b]
R-411A	zeotrope	R-1270/22/152a (1.5/87.5/11.0)	A2	2.9	14,000	46	990	—
R-411B	zeotrope	R-1270/22/152a (3.0/94.0/3.0)	A2	2.8	13,000	45	980	—
R-412A	zeotrope	R-22/218/142b (70.0/5.0/25.0)	A2	5.1	22,000	82	1,000	—
R-413A	zeotrope	R-218/134a/600a (9.0/88.0/3.0)	A2	5.8	22,000	94	1,000	—
R-414A	zeotrope	R-22/124/600a/142b (51.0/28.5/4.0/16.5)	A1	6.4	26,000	100	1,000	—
R-414B	zeotrope	R-22/124/600a/142b (50.0/39.0/1.5/9.5)	A1	6.0	23,000	95	1,000	—
R-415A	zeotrope	R-22/152a (82.0/18.0)	A2	2.9	14,000	47	1,000	—
R-415B	zeotrope	R-22/152a (25.0/75.0)	A2	2.1	12,000	34	1,000	—
R-416A	zeotrope	R-134a/124/600 (59.0/39.5/1.5)	A1	3.9	14,000	62	1,000	2-0-0[b]
R-417A	zeotrope	R-125/134a/600 (46.6/50.0/3.4)	A1	3.5	13,000	56	1,000	2-0-0[b]
R-417B	zeotrope	R-125/134a/600 (79.0/18.3/2.7)	A1	4.3	15,000	70	1,000	—
R-418A	zeotrope	R-290/22/152a (1.5/96.0/2.5)	A2	4.8	22,000	77	1,000	—

(continued)

TABLE 1103.1—continued
REFRIGERANT CLASSIFICATION, AMOUNT AND OEL

CHEMICAL REFRIGERANT	FORMULA	CHEMICAL NAME OF BLEND	REFRIGERANT CLASSIFICATION	AMOUNT OF REFRIGERANT PER OCCUPIED SPACE				[F] DEGREES OF HAZARD[a]
				Pounds per 1,000 cubic feet	ppm	g/m³	OEL[e]	
R-419A	zeotrope	R-125/134a/E170 (77.0/19.0/4.0)	A2	4.2	15,000	67	1,000	—
R-420A	zeotrope	R-134a/142b (88.0/12.0)	A1	12	45,000	190	1,000	2-0-0[b]
R-421A	zeotrope	R-125/134a (58.0/42.0)	A1	17	61,000	280	1,000	2-0-0[b]
R-421B	zeotrope	R-125/134a (85.0/15.0)	A1	21	69,000	330	1,000	2-0-0[b]
R-422A	zeotrope	R-125/134a/600a (85.1/11.5/3.4)	A1	18	63,000	290	1,000	2-0-0[b]
R-422B	zeotrope	R-125/134a/600a (55.0/42.0/3.0)	A1	16	56,000	250	1,000	2-0-0[b]
R-422C	zeotrope	R-125/134a/600a (82.0/15.0/3.0)	A1	18	62,000	290	1,000	2-0-0[b]
R-422D	zeotrope	R-125/134a/600a (65.1/31.5/3.4)	A1	16	58,000	260	1,000	2-0-0[b]
R-423A	zeotrope	R-134a/227ea (52.5/47.5)	A1	19	59,000	310	1,000	2-0-0[c]
R-424A	zeotrope	R-125/134a/600a/600/601a (50.5/47.0/0.9/1.0/0.6)	A1	6.2	23,000	100	970	2-0-0[b]
R-425A	zoetrope	R-32/134a/227ea (18.5/69.5/12.0)	A1	16	72,000	260	1,000	2-0-0[b]
R-426A	zeotrope	R-125/134a/600a/601a (5.1/93.0/1.3/0.6)	A1	5.2	20,000	83	990	—
R-427A	zeotrope	R-32/125/143a/134a (15.0/25.0/10.0/50.0)	A1	18	79,000	290	1,000	2-1-0
R-428A	zeotrope	R-125/143a/290/600a (77.5/20.0/0.6/1.9)	A1	23	83,000	370	1,000	—
R-429A	zeotrope	R-E170/152a/600a (60.0/10.0/30.0)	A3	0.81	6,300	13	1,000	—
R-430A	zeotrope	R-152a/600a (76.0/24.0)	A3	1.3	8,000	21	1,000	—
R-431A	zeotrope	R-290/152a (71.0/29.0)	A3	0.69	5,500	11	1,000	—
R-432A	zeotrope	R-1270/E170 (80.0/20.0)	A3	0.13	1,200	2.1	700	—
R-433A	zeotrope	R-1270/290 (30.0/70.0)	A3	0.34	3,100	5.5	880	—
R-433B	zeotrope	R-1270/290 (5.0-95.0)	A3	0.51	4,500	8.1	950	—
R-433C	zeotrope	R-1270/290 (25.0-75.0)	A3	0.41	3,600	6.6	790	—
R-434A	zeotrope	R-125/143a/600a (63.2/18.0/16.0/2.8)	A1	20	73,000	320	1,000	—
R-435A	zeotrope	R-E170/152a (80.0/20.0)	A3	1.1	8,500	17	1,000	—
R-436A	zeotrope	R-290/600a (56.0/44.0)	A3	0.50	4,000	8.1	1,000	—
R-436B	zeotrope	R-290/600a (52.0/48.0)	A3	0.51	4,000	8.1	1,000	—
R-437A	zeotrope	R-125/134a/600/601 (19.5/78.5/1.4/0.6)	A1	5.0	19,000	82	990	—
R-438A	zeotrope	R-32/125/134a/600/601a (8.5/45.0/44.2/1.7/0.6)	A1	4.9	20,000	79	990	—
R-439A	zeotrope	R-32/125/600a (50.0/47.0/3.0)	A2	4.7	26,000	76	990	—
R-440A	zeotrope	R-290/134a/152a (0.6/1.6/97.8)	A2	1.9	12,000	31	1,000	—
R-441A	zeotrope	R-170/290/600a/600 (3.1/54.8/6.0/36.1)	A3	0.39	3,200	6.3	1,000	—
R-442A	zeotrope	R-32/125/134a/152a/227ea (31.0/31.0/30.0/3.0/5.0)	A1	21	100,000	330	1,000	—

(continued)

TABLE 1103.1—continued
REFRIGERANT CLASSIFICATION, AMOUNT AND OEL

CHEMICAL REFRIGERANT	FORMULA	CHEMICAL NAME OF BLEND	REFRIGERANT CLASSIFICATION	AMOUNT OF REFRIGERANT PER OCCUPIED SPACE				[F] DEGREES OF HAZARD[a]
				Pounds per 1,000 cubic feet	ppm	g/m³	OEL[e]	
R-500[e]	azeotrope	R-12/152a (73.8/26.2)	A1	7.6	30,000	120	1,000	2-0-0[b]
R-501[d]	azeotrope	R-22/12 (75.0/25.0)	A1	13	54,000	210	1,000	—
R-502[e]	azeotrope	R-22/115 (48.8/51.2)	A1	21	73,000	330	1,000	2-0-0[b]
R-503[e]	azeotrope	R-23/13 (40.1/59.9)	—	—	—	—	1,000	2-0-0[b]
R-504[d]	azeotrope	R-32/115 (48.2/51.8)	—	28	140,000	450	1,000	2-0-0[b]
R-507A	azeotrope	R-125/143a (50.0/50.0)	A1	32	130,000	520	1,000	2-0-0[b]
R-508A	azeotrope	R-23/116 (39.0/61.0)	A1	14	55,000	220	1,000	2-0-0[b]
R-508B	azeotrope	R-23/116 (46.0/54.0)	A1	13	52,000	200	1,000	2-0-0[b]
R-509A	azeotrope	R-22/218 (44.0/56.0)	A1	24	75,000	390	1,000	2-0-0[b]
R-510A	azeotrope	R-E170/600a (88.0/12.0)	A3	0.87	7,300	14	1,000	—
R-511A	azeotrope	R-290/E170 (95.0/5.0)	A3	0.59	5,300	9.5	1,000	—
R-512A	azeotrope	R-134a/152a (5.0/95.0)	A2	1.9	11,000	31	1,000	—
R-600	$CH_3CH_2CH_2CH_3$	butane	A3	0.15	1,000	2.4	1,000	1-4-0
R-600a	$CH(CH_3)_2CH_3$	2-methylpropane (isobutane)	A3	0.59	4,000	9.6	1,000	2-4-0
R-601	$CH_3CH_2CH_2CH_2CH_3$	pentane	A3	0.18	1,000	2.9	600	—
R-601a	$(CH_3)_2CHCH_2CH_3$	2-methylbutane (isopentane)	A3	0.18	1,000	2.9	600	—
R-717	NH_3	ammonia	B2[f]	0.014	320	0.22	25	3-3-0[c]
R-718	H_2O	water	A1	—	—	—	—	0-0-0
R-744	CO_2	carbon dioxide	A1	4.5	40,000	72	5,000	2-0-0[b]
R-1150	$CH_2=CH_2$	ethene (ethylene)	A3	—	—	—	200	1-4-2
R-1234yf	$CF_3CF=CH_2$	2,3,3,3-tetrafluoro-1 propene	A2[f]	4.7	16,000	75	500	—
R-1234ze(E)	$CF3CH=CHF$	trans-1,3,3,3-tetrafluoro-1-propene	A2[f]	4.7	16,000	75	800	—
R-1270	$CH_3CH=CH_2$	Propene (propylene)	A3	0.1	1,000	1.7	500	1-4-1

For SI: 1 pound = 0.454 kg, 1 cubic foot = 0.0283 m³.

a. Degrees of hazard are for health, fire, and reactivity, respectively, in accordance with NFPA 704.
b. Reduction to 1-0-0 is allowed if analysis satisfactory to the code official shows that the maximum concentration for a rupture or full loss of refrigerant charge would not exceed the IDLH, considering both the refrigerant quantity and room volume.
c. For installations that are entirely outdoors, use 3-1-0.
d. Class I ozone depleting substance; prohibited for new installations.
e. Occupational Exposure Limit based on the OSHA PEL, ACGIH TLV-TWA, the AIHA WEEL or consistent value on a time-weighted average (TWA) basis (unless noted C for ceiling) for an 8 hr/d and 40 hr/wk.
f. The ASHRAE Standard 34 flammability classification for this refrigerant is 2L, which is a subclass of Class 2.

SECTION 1104
SYSTEM APPLICATION REQUIREMENTS

1104.1 General. The refrigerant, occupancy and system classification cited in this section shall be determined in accordance with Sections 1103.1, 1103.2 and 1103.3, respectively.

1104.2 Machinery room. Except as provided in Sections 1104.2.1 and 1104.2.2, all components containing the refrigerant shall be located either outdoors or in a *machinery room* where the quantity of refrigerant in an independent circuit of a system exceeds the amounts shown in Table 1103.1. For refrigerant blends not listed in Table 1103.1, the same requirement shall apply when the amount for any blend component exceeds that indicated in Table 1103.1 for that component. This requirement shall also apply when the combined amount of the blend components exceeds a limit of 69,100 parts per million (ppm) by volume. Machinery rooms required by this section shall be constructed and maintained in accordance with Section 1105 for Group A1 and B1 refrigerants and in accordance with Sections 1105 and 1106 for Group A2, B2, A3 and B3 refrigerants.

Exceptions:

1. Machinery rooms are not required for *listed equipment* and appliances containing not more than 6.6 pounds (3 kg) of refrigerant, regardless of the refrigerant's safety classification, where installed in accordance with the equipment's or appliance's listing and the *equipment* or *appliance* manufacturer's installation instructions.

2. Piping in conformance with Section 1107 is allowed in other locations to connect components installed in a *machinery room* with those installed outdoors.

1104.2.1 Institutional occupancies. The amounts shown in Table 1103.1 shall be reduced by 50 percent for all areas of institutional occupancies except kitchens, laboratories and mortuaries. The total of all Group A2, B2, A3 and B3 refrigerants shall not exceed 550 pounds (250 kg) in occupied areas or machinery rooms.

1104.2.2 Industrial occupancies and refrigerated rooms. This section applies only to industrial occupancies and refrigerated rooms for manufacturing, food and beverage preparation, meat cutting, other processes and storage. Machinery rooms are not required where all of the following conditions are met:

1. The space containing the machinery is separated from other occupancies by tight construction with tight-fitting doors.

2. Access is restricted to authorized personnel.

3. The floor area per occupant is not less than 100 square feet (9.3 m^2) where machinery is located on floor levels with exits more than 6.6 feet (2012 mm) above the ground. Where provided with egress directly to the outdoors or into *approved* building exits, the minimum floor area shall not apply.

4. Refrigerant detectors are installed as required for machinery rooms in accordance with Section 1105.3.

5. Surfaces having temperatures exceeding 800°F (427°C) and open flames are not present where any Group A2, B2, A3 or B3 refrigerant is used (see Section 1104.3.4).

6. All electrical *equipment* and appliances conform to Class 1, Division 2, *hazardous location* classification requirements of NFPA 70 where the quantity of any Group A2, B2, A3 or B3 refrigerant, other than ammonia, in a single independent circuit would exceed 25 percent of the lower flammability limit (LFL) upon release to the space.

7. All refrigerant-containing parts in systems exceeding 100 horsepower (hp) (74.6 kW) drive power, except evaporators used for refrigeration or dehumidification; condensers used for heating; control and pressure relief valves for either; and connecting piping, shall be located either outdoors or in a *machinery room*.

1104.3 Refrigerant restrictions. Refrigerant applications, maximum quantities and use shall be restricted in accordance with Sections 1104.3.1 through 1104.3.4.

1104.3.1 Air-conditioning for human comfort. In other than industrial occupancies where the quantity in a single independent circuit does not exceed the amount in Table 1103.1, Group B1, B2 and B3 refrigerants shall not be used in high-probability systems for air-conditioning for human comfort.

1104.3.2 Nonindustrial occupancies. Group A2 and B2 refrigerants shall not be used in high-probability systems where the quantity of refrigerant in any independent refrigerant circuit exceeds the amount shown in Table 1104.3.2. Group A3 and B3 refrigerants shall not be used except where *approved*.

Exception: This section does not apply to laboratories where the floor area per occupant is not less than 100 square feet (9.3 m^2).

1104.3.3 All occupancies. The total of all Group A2, B2, A3 and B3 refrigerants other than R-717, ammonia, shall not exceed 1,100 pounds (499 kg) except where *approved*.

1104.3.4 Protection from refrigerant decomposition. Where any device having an open flame or surface temperature greater than 800°F (427°C) is used in a room containing more than 6.6 pounds (3 kg) of refrigerant in a single independent circuit, a hood and exhaust system shall be provided in accordance with Section 510. Such exhaust system shall exhaust *combustion* products to the outdoors.

Exception: A hood and exhaust system shall not be required where any of the following apply:

1. The refrigerant is R-717, R-718 or R-744.

2. The *combustion* air is ducted from the outdoors in a manner that prevents leaked refrigerant from being combusted.

3. A refrigerant detector is used to stop the *combustion* in the event of a refrigerant leak (see Sections 1105.3 and 1105.5).

TABLE 1104.3.2
MAXIMUM PERMISSIBLE QUANTITIES OF REFRIGERANTS

TYPE OF REFRIGERATION SYSTEM	MAXIMUM POUNDS FOR VARIOUS OCCUPANCIES			
	Institutional	Assembly	Residential	All other occupancies
Sealed absorption system				
In exit access	0	0	3.3	3.3
In adjacent outdoor locations	0	0	22	22
In other than exit access	0	6.6	6.6	6.6
Unit systems				
In other than exit access	0	0	6.6	6.6

For SI: 1 pound = 0.454 kg.

1104.4 Volume calculations. Volume calculations shall be in accordance with Sections 1104.4.1 through 1104.4.3.

1104.4.1 Noncommunicating spaces. Where the refrigerant-containing parts of a system are located in one or more spaces that do not communicate through permanent openings or HVAC ducts, the volume of the smallest, enclosed occupied space shall be used to determine the permissible quantity of refrigerant in the system.

1104.4.2 Communicating spaces. Where an evaporator or condenser is located in an air duct system, the volume of the smallest, enclosed occupied space served by the duct system shall be used to determine the maximum allowable quantity of refrigerant in the system.

Exception: If airflow to any enclosed space cannot be reduced below one-quarter of its maximum, the entire space served by the air duct system shall be used to determine the maximum allowable quantity of refrigerant in the system.

1104.4.3 Plenums. Where the space above a suspended ceiling is continuous and part of the supply or return air *plenum* system, this space shall be included in calculating the volume of the enclosed space.

SECTION 1105
MACHINERY ROOM, GENERAL REQUIREMENTS

[BF] 1105.1 Design and construction. Machinery rooms shall be designed and constructed in accordance with the *International Building Code* and this section.

1105.2 Openings. Ducts and air handlers in the *machinery room* that operate at a lower pressure than the room shall be sealed to prevent any refrigerant leakage from entering the airstream.

[F] 1105.3 Refrigerant detector. Refrigerant detectors in machinery rooms shall be provided as required by Section 606.8 of the *International Fire Code*.

1105.4 Tests. Periodic tests of the mechanical ventilating system shall be performed in accordance with manufacturer's specifications and as required by the code official.

1105.5 Fuel-burning appliances. Fuel-burning appliances and *equipment* having open flames and that use *combustion*

air from the *machinery room* shall not be installed in a *machinery room*.

Exceptions:

1. Where the refrigerant is carbon dioxide or water.

2. Fuel-burning appliances shall not be prohibited in the same *machinery room* with refrigerant-containing *equipment* or appliances where *combustion* air is ducted from outside the *machinery room* and sealed in such a manner as to prevent any refrigerant leakage from entering the *combustion* chamber, or where a refrigerant vapor detector is employed to automatically shut off the *combustion* process in the event of refrigerant leakage.

1105.6 Ventilation. Machinery rooms shall be mechanically ventilated to the outdoors.

Exception: Where a refrigerating system is located outdoors more than 20 feet (6096 mm) from any building opening and is enclosed by a penthouse, lean-to or other open structure, natural or mechanical ventilation shall be provided. Location of the openings shall be based on the relative density of the refrigerant to air. The free-aperture cross section for the ventilation of the *machinery room* shall be not less than:

$$F = \sqrt{G} \qquad \text{(Equation 11-1)}$$

For SI: $F = 0.138\sqrt{G}$

where:

F = The free opening area in square feet (m²).

G = The mass of refrigerant in pounds (kg) in the largest system, any part of which is located in the *machinery room*.

1105.6.1 Discharge location. The discharge of the air shall be to the outdoors in accordance with Chapter 5. Exhaust from mechanical ventilation systems shall be discharged not less than 20 feet (6096 mm) from a property line or openings into buildings.

1105.6.2 Makeup air. Provisions shall be made for *makeup air* to replace that being exhausted. Openings for *makeup air* shall be located to avoid intake of *exhaust air*. Supply and exhaust ducts to the *machinery room* shall not serve any other area, shall be constructed in accordance with Chapter 5 and shall be covered with corrosion-resistant screen of not less than $^1/_4$-inch (6.4 mm) mesh.

1105.6.3 Ventilation rate. For other than ammonia systems, the mechanical ventilation systems shall be capable of exhausting the minimum quantity of air both at normal operating and emergency conditions, as required by Sections 1105.6.3.1 and 1105.6.3.2. The minimum required ventilation rate for ammonia shall be 30 air changes per hour in accordance with IIAR2. Multiple fans or multi-speed fans shall be allowed to produce the emergency ventilation rate and to obtain a reduced airflow for normal ventilation.

1105.6.3.1 Quantity—normal ventilation. During occupied conditions, the mechanical ventilation system shall exhaust the larger of the following:

1. Not less than 0.5 cfm per square foot (0.0025 m³/s • m²) of *machinery room* area or 20 cfm (0.009 m³/s) per person.

2. A volume required to limit the room temperature rise to 18°F (10°C) taking into account the ambient heating effect of all machinery in the room.

1105.6.3.2 Quantity—emergency conditions. Upon actuation of the refrigerant detector required in Section 1105.3, the mechanical ventilation system shall *exhaust air* from the *machinery room* in the following quantity:

$$Q = 100 \times \sqrt{G} \qquad \text{(Equation 11-2)}$$

For SI: $Q = 0.07 \times \sqrt{G}$

where:

Q = The airflow in cubic feet per minute (m³/s).

G = The design mass of refrigerant in pounds (kg) in the largest system, any part of which is located in the *machinery room.*

1105.7 Termination of relief devices. Pressure relief devices, fusible plugs and purge systems located within the *machinery room* shall terminate outside of the structure at a location not less than 15 feet (4572 mm) above the adjoining grade level and not less than 20 feet (6096 mm) from any window, ventilation opening or exit.

1105.8 Ammonia discharge. Pressure relief valves for ammonia systems shall discharge in accordance with ASHRAE 15.

[F] 1105.9 Emergency pressure control system. Permanently installed refrigeration systems containing more than 6.6 pounds (3 kg) of flammable, toxic or highly toxic refrigerant or ammonia shall be provided with an emergency pressure control system in accordance with Section 606.10 of the *International Fire Code.*

SECTION 1106
MACHINERY ROOM, SPECIAL REQUIREMENTS

1106.1 General. Where required by Section 1104.2, the *machinery room* shall meet the requirements of this section in addition to the requirements of Section 1105.

1106.2 Elevated temperature. There shall not be an open flame-producing device or continuously operating hot surface over 800°F (427°C) permanently installed in the room.

1106.3 Ammonia room ventilation. Ventilation systems in ammonia machinery rooms shall be operated continuously at the ventilation rate specified in Section 1105.6.3.

Exceptions:

1. Machinery rooms equipped with a vapor detector that will automatically start the ventilation system at the ventilation rate specified in Section 1105.6.3, and that will actuate an alarm at a detection level not to exceed 1,000 ppm.

2. Machinery rooms conforming to the Class 1, Division 2, *hazardous location* classification requirements of NFPA 70.

1106.4 Flammable refrigerants. Where refrigerants of Groups A2, A3, B2 and B3 are used, the *machinery room* shall conform to the Class 1, Division 2, *hazardous location* classification requirements of NFPA 70.

Exception: Ammonia machinery rooms that are provided with ventilation in accordance with Section 1106.3.

[F] 1106.5 Remote controls. Remote control of the mechanical equipment and appliances located in the machinery room shall comply with Sections 1106.5.1 and 1106.5.2.

[F] 1106.5.1 Refrigeration system emergency shutoff. A clearly identified switch of the break-glass type or with an approved tamper-resistant cover shall provide off-only control of refrigerant compressors, refrigerant pumps, and normally closed, automatic refrigerant valves located in the machinery room. Additionally, this equipment shall be automatically shut off whenever the refrigerant vapor concentration in the machinery room exceeds the vapor detector's upper detection limit or 25 percent of the LEL, whichever is lower.

[F] 1106.5.2 Ventilation system. A clearly identified switch of the break-glass type or with an approved tamper-resistant cover shall provide on-only control of the *machinery room* ventilation fans.

[F] 1106.6 Emergency signs and labels. Refrigeration units and systems shall be provided with *approved* emergency signs, charts, and labels in accordance with the *International Fire Code.*

SECTION 1107
REFRIGERANT PIPING

1107.1 General. The design of refrigerant piping shall be in accordance with ASME B31.5. Refrigerant piping shall be installed, tested and placed in operation in accordance with this chapter.

1107.2 Piping location. Refrigerant piping that crosses an open space that affords passageway in any building shall be not less than 7 feet 3 inches (2210 mm) above the floor unless the piping is located against the ceiling of such space. Refrigerant piping shall not be placed in any elevator, dumbwaiter or other shaft containing a moving object or in any shaft that has openings to living quarters or to means of egress. Refrigerant piping shall not be installed in an enclosed public stairway, stairway landing or means of egress.

1107.2.1 Piping in concrete floors. Refrigerant piping installed in concrete floors shall be encased in pipe ducts. The piping shall be isolated and supported to prevent damaging vibration, stress and corrosion.

1107.2.2 Refrigerant penetrations. Refrigerant piping shall not penetrate floors, ceilings or roofs.

Exceptions:

1. Penetrations connecting the basement and the first floor.

2. Penetrations connecting the top floor and a machinery penthouse or roof installation.

3. Penetrations connecting adjacent floors served by the refrigeration system.

4. Penetrations by piping in a direct system where the refrigerant quantity does not exceed Table 1103.1 for the smallest occupied space through which the piping passes.

5. In other than industrial occupancies and where the refrigerant quantity exceeds Table 1103.1 for the smallest space, penetrations for piping that connects separate pieces of *equipment* that are either:

    5.1. Enclosed by an *approved* gas-tight, fire-resistive duct or shaft with openings to those floors served by the refrigeration system.

    5.2. Located on the exterior of the building where vented to the outdoors or to the space served by the system and not used as an air shaft, closed court or similar space.

1107.3 Pipe enclosures. Rigid or flexible metal enclosures or pipe ducts shall be provided for soft, annealed copper tubing and used for refrigerant piping erected on the premises and containing other than Group A1 or B1 refrigerants. Enclosures shall not be required for connections between condensing units and the nearest riser box(es), provided such connections do not exceed 6 feet (1829 mm) in length.

1107.4 Condensation. Refrigerating piping and fittings, brine piping and fittings that, during normal operation, will reach a surface temperature below the dew point of the surrounding air, and are located in spaces or areas where condensation will cause a safety hazard to the building occupants, structure, electrical *equipment* or any other *equipment* or appliances, shall be protected in an *approved* manner to prevent such damage.

1107.5 Materials for refrigerant pipe and tubing. Piping materials shall be as set forth in Sections 1107.5.1 through 1107.5.5.

1107.5.1 Steel pipe. Carbon steel pipe with a wall thickness not less than Schedule 80 shall be used for Group A2, A3, B2 or B3 refrigerant liquid lines for sizes 1.5 inches (38 mm) and smaller. Carbon steel pipe with a wall thickness not less than Schedule 40 shall be used for Group A1 or B1 refrigerant liquid lines 6 inches (152 mm) and smaller, Group A2, A3, B2 or B3 refrigerant liquid lines sizes 2 inches (51 mm) through 6 inches (152 mm) and all refrigerant suction and discharge lines 6 inches (152 mm) and smaller. Type F steel pipe shall not be used for refrigerant lines having an operating temperature less than -20°F (-29°C).

1107.5.2 Copper and brass pipe. Standard iron-pipe size, copper and red brass (not less than 80-percent copper) pipe shall conform to ASTM B 42 and ASTM B 43.

1107.5.3 Copper tube. Copper tube used for refrigerant piping erected on the premises shall be seamless copper tube of Type ACR (hard or annealed) complying with ASTM B 280. Where *approved*, copper tube for refrigerant piping erected on the premises shall be seamless copper tube of Type K, L or M (drawn or annealed) in accordance with ASTM B 88. Annealed temper copper tube shall not be used in sizes larger than a 2-inch (51 mm) nominal size. Mechanical joints shall not be used on annealed temper copper tube in sizes larger than $^{7}/_{8}$-inch (22.2 mm) OD size.

1107.5.4 Copper tubing joints. Copper tubing joints used in refrigerating systems containing Group A2, A3, B2 or B3 refrigerants shall be brazed. Soldered joints shall not be used in such refrigerating systems.

1107.5.5 Aluminum tube. Type 3003-0 aluminum tubing with high-pressure fittings shall not be used with methyl chloride and other refrigerants known to attack aluminum.

1107.6 Joints and refrigerant-containing parts in air ducts. Joints and all refrigerant-containing parts of a refrigerating system located in an air duct of an air-conditioning system carrying conditioned air to and from human-occupied space shall be constructed to withstand, without leakage, a pressure of 150 percent of the higher of the design pressure or pressure relief device setting.

1107.7 Exposure of refrigerant pipe joints. Refrigerant pipe joints erected on the premises shall be exposed for visual inspection prior to being covered or enclosed.

1107.8 Stop valves. Systems containing more than 6.6 pounds (3 kg) of a refrigerant in systems using positive-displacement compressors shall have stop valves installed as follows:

1. At the inlet of each compressor, compressor unit or condensing unit.

2. At the discharge outlet of each compressor, compressor unit or condensing unit and of each liquid receiver.

Exceptions:

1. Systems that have a refrigerant pumpout function capable of storing the entire refrigerant charge in a receiver or heat exchanger.

2. Systems that are equipped with provisions for pumpout of the refrigerant using either portable or permanently installed recovery *equipment.*

3. Self-contained systems.

1107.8.1 Liquid receivers. Systems containing 100 pounds (45 kg) or more of a refrigerant, other than systems utilizing nonpositive displacement compressors, shall have stop valves, in addition to those required by Section 1107.8, on each inlet of each liquid receiver. Stop valves shall not be required on the inlet of a receiver in a condensing unit, nor on the inlet of a receiver which is an integral part of the condenser.

1107.8.2 Copper tubing. Stop valves used with soft annealed copper tubing or hard-drawn copper tubing $^7/_8$-inch (22.2 mm) OD standard size or smaller shall be securely mounted, independent of tubing fastenings or supports.

1107.8.3 Identification. Stop valves shall be identified where their intended purpose is not obvious. Numbers shall not be used to label the valves, unless a key to the numbers is located near the valves.

SECTION 1108
FIELD TEST

1108.1 General. Every refrigerant-containing part of every system that is erected on the premises, except compressors, condensers, vessels, evaporators, safety devices, pressure gauges and control mechanisms that are *listed* and factory tested, shall be tested and proved tight after complete installation, and before operation. Tests shall include both the high- and low-pressure sides of each system at not less than the lower of the design pressures or the setting of the pressure relief device(s). The design pressures for testing shall be those listed on the condensing unit, compressor or compressor unit nameplate, as required by ASHRAE 15.

Exceptions:

1. Gas bulk storage tanks that are not permanently connected to a refrigeration system.

2. Systems erected on the premises with copper tubing not exceeding $^5/_8$-inch (15.8 mm) OD, with wall thickness as required by ASHRAE 15, shall be tested in accordance with Section 1108.1, or by means of refrigerant charged into the system at the saturated vapor pressure of the refrigerant at 70°F (21°C) or higher.

3. Limited-charge systems equipped with a pressure relief device, erected on the premises, shall be tested at a pressure not less than one and one-half times the pressure setting of the relief device. If the *equipment* or *appliance* has been tested by the manufacturer at one and one-half times the design pressure, the test after erection on the premises shall be conducted at the design pressure.

1108.1.1 Booster compressor. Where a compressor is used as a booster to obtain an intermediate pressure and discharges into the suction side of another compressor, the booster compressor shall be considered a part of the low side, provided that it is protected by a pressure relief device.

1108.1.2 Centrifugal/nonpositive displacement compressors. In field-testing systems using centrifugal or other nonpositive displacement compressors, the entire system shall be considered as the low-side pressure for field test purposes.

1108.2 Test gases. Tests shall be performed with an inert dried gas including, but not limited to, nitrogen and carbon dioxide. Oxygen, air, combustible gases and mixtures containing such gases shall not be used.

Exception: The use of air is allowed to test R-717, ammonia, systems provided that they are subsequently evacuated before charging with refrigerant.

1108.3 Test apparatus. The means used to build up the test pressure shall have either a pressure-limiting device or a pressure-reducing device and a gauge on the outlet side.

1108.4 Declaration. A certificate of test shall be provided for all systems containing 55 pounds (25 kg) or more of refrigerant. The certificate shall give the name of the refrigerant and the field test pressure applied to the high side and the low side of the system. The certification of test shall be signed by the installer and shall be made part of the public record.

[F] SECTION 1109
PERIODIC TESTING

[F] 1109.1 Testing required. The following emergency devices and systems shall be periodically tested in accordance with the manufacturer's instructions and as required by the code official:

1. Treatment and flaring systems.

2. Valves and appurtenances necessary to the operation of emergency refrigeration control boxes.

3. Fans and associated *equipment* intended to operate emergency ventilation systems.

4. Detection and alarm systems.

CHAPTER 12

HYDRONIC PIPING

SECTION 1201
GENERAL

1201.1 Scope. The provisions of this chapter shall govern the construction, installation, *alteration* and repair of hydronic piping systems. This chapter shall apply to hydronic piping systems that are part of heating, ventilation and air-conditioning systems. Such piping systems shall include steam, hot water, chilled water, steam condensate and ground source heat pump loop systems. Potable cold and hot water distribution systems shall be installed in accordance with the *International Plumbing Code*.

1201.2 Sizing. Piping and piping system components for hydronic systems shall be sized for the demand of the system.

1201.3 Standards. As an alternative to the provisions of Sections 1202 and 1203, piping shall be designed, installed, inspected and tested in accordance with ASME B31.9.

SECTION 1202
MATERIAL

1202.1 Piping. Piping material shall conform to the standards cited in this section.

Exception: Embedded piping regulated by Section 1209.

1202.2 Used materials. Reused pipe, fittings, valves or other materials shall be clean and free of foreign materials and shall be *approved* by the code official for reuse.

1202.3 Material rating. Materials shall be rated for the operating temperature and pressure of the hydronic system. Materials shall be suitable for the type of fluid in the hydronic system.

1202.4 Piping materials standards. Hydronic pipe shall conform to the standards listed in Table 1202.4. The exterior of the pipe shall be protected from corrosion and degradation.

TABLE 1202.4
HYDRONIC PIPE

MATERIAL	STANDARD (see Chapter 15)
Acrylonitrile butadiene styrene (ABS) plastic pipe	ASTM D 1527; ASTM F 2806
Chlorinated polyvinyl chloride (CPVC) plastic pipe	ASTM D 2846; ASTM F 441; ASTM F 442
Copper or copper-alloy pipe	ASTM B 42; ASTM B 43; ASTM B 302
Copper or copper-alloy tube (Type K, L or M)	ASTM B 75; ASTM B 88; ASTM B 135; ASTM B 251
Cross-linked polyethylene/ aluminum/cross-linked polyethylene (PEX-AL-PEX) pressure pipe	ASTM F 1281; CSA CAN/CSA-B-137.10

(continued)

TABLE 1202.4—continued
HYDRONIC PIPE

MATERIAL	STANDARD (see Chapter 15)
Cross-linked polyethylene (PEX) tubing	ASTM F 876; ASTM F 877
Ductile iron pipe	AWWA C115/A21.15; AWWA C151/A21.51
Lead pipe	FS WW-P-325B
Polyethylene/aluminum/polyethylene (PE-AL-PE) pressure pipe	ASTM F 1282; CSA B137.9
Polypropylene (PP) plastic pipe	ASTM F 2389
Polyvinyl chloride (PVC) plastic pipe	ASTM D 1785; ASTM D 2241
Raised temperature polyethylene (PE-RT)	ASTM F 2623; ASTM F 2769
Steel pipe	ASTM A 53; ASTM A 106
Steel tubing	ASTM A 254

1202.5 Pipe fittings. Hydronic pipe fittings shall be *approved* for installation with the piping materials to be installed, and shall conform to the respective pipe standards or to the standards listed in Table 1202.5.

TABLE 1202.5
HYDRONIC PIPE FITTINGS

MATERIAL	STANDARD (see Chapter 15)
Copper and copper alloys	ASME B16.15; ASME B16.18; ASME B16.22; ASME B16.26; ASTM F 1974; ASTM B16.24; ASME B16.51
Ductile iron and gray iron	ANSI/AWWA C110/A21.10; AWWA C153/A21.53; ASTM A 395; ASTM A 536; ASTM F 1476; ASTM F 1548
Ductile iron	ANSI/AWWA C153/A21.53
Gray iron	ASTM A 126
Malleable iron	ASME B16.3
PE-RT fittings	ASTM F 1807; ASTM F 2098; ASTM F 2159; ASTM F 2735; ASTM F 2769
PEX fittings	ASTM F 877; ASTM F 1807; ASTM F 2159
Plastic	ASTM D 2466; ASTM D 2467; ASTM F 438; ASTM F 439; ASTM F 877; ASTM F 2389; ASTM F 2735
Steel	ASME B16.5; ASME B16.9; ASME B16.11; ASME B16.28; ASTM A 53; ASTM A 106; ASTM A 234; ASTM A 420; ASTM A 536; ASTM A 395; ASTM F 1476; ASTM F 1548

1202.6 Valves. Valves shall be constructed of materials that are compatible with the type of piping material and fluids in the system. Valves shall be rated for the temperatures and pressures of the systems in which the valves are installed.

1202.7 Flexible connectors, expansion and vibration compensators. Flexible connectors, expansion and vibration control devices and fittings shall be of an *approved* type.

SECTION 1203
JOINTS AND CONNECTIONS

1203.1 Approval. Joints and connections shall be of an *approved* type. Joints and connections shall be tight for the pressure of the hydronic system.

1203.1.1 Joints between different piping materials. Joints between different piping materials shall be made with *approved* adapter fittings.

1203.2 Preparation of pipe ends. Pipe shall be cut square, reamed and chamfered, and shall be free of burrs and obstructions. Pipe ends shall have full-bore openings and shall not be undercut.

1203.3 Joint preparation and installation. Where required by Sections 1203.4 through 1203.14, the preparation and installation of brazed, mechanical, soldered, solvent-cemented, threaded and welded joints shall comply with Sections 1203.3.1 through 1203.3.8.

1203.3.1 Brazed joints. Joint surfaces shall be cleaned. An *approved* flux shall be applied where required. The joint shall be brazed with a filler metal conforming to AWS A5.8.

1203.3.2 Mechanical joints. Mechanical joints shall be installed in accordance with the manufacturer's instructions.

1203.3.3 Soldered joints. Joint surfaces shall be cleaned. A flux conforming to ASTM B 813 shall be applied. The joint shall be soldered with a solder conforming to ASTM B 32.

1203.3.4 Solvent-cemented joints. Joint surfaces shall be clean and free of moisture. An *approved* primer shall be applied to CPVC and PVC pipe-joint surfaces. Joints shall be made while the cement is wet. Solvent cement conforming to the following standards shall be applied to all joint surfaces:

1. ASTM D 2235 for ABS joints.

2. ASTM F 493 for CPVC joints.

3. ASTM D 2564 for PVC joints.

CPVC joints shall be made in accordance with ASTM D 2846.

Exception: For CPVC pipe joint connections, a primer is not required where all of the following conditions apply:

1. The solvent cement used is third-party certified as conforming to ASTM F 493.

2. The solvent cement is yellow in color.

3. The solvent cement is used only for joining $^1/_2$-inch (12.7 mm) through 2-inch (51 mm) diameter CPVC pipe and fittings.

4. The CPVC pipe and fittings are manufactured in accordance with ASTM D 2846.

1203.3.5 Threaded joints. Threads shall conform to ASME B1.20.1. Schedule 80 or heavier plastic pipe shall be threaded with dies specifically designed for plastic pipe. Thread lubricant, pipe-joint compound or tape shall be applied on the male threads only and shall be *approved* for application on the piping material.

1203.3.6 Welded joints. Joint surfaces shall be cleaned by an *approved* procedure. Joints shall be welded with an *approved* filler metal.

1203.3.7 Grooved and shouldered mechanical joints. Grooved and shouldered mechanical joints shall conform to the requirements of ASTM F 1476 and shall be installed in accordance with the manufacturer's instructions.

1203.3.8 Mechanically formed tee fittings. Mechanically extracted outlets shall have a height not less than three times the thickness of the branch tube wall.

1203.3.8.1 Full flow assurance. Branch tubes shall not restrict the flow in the run tube. A dimple/depth stop shall be formed in the branch tube to ensure that penetration into the outlet is of the correct depth. For inspection purposes, a second dimple shall be placed $^1/_4$ inch (6.4 mm) above the first dimple. Dimples shall be aligned with the tube run.

1203.3.8.2 Brazed joints. Mechanically formed tee fittings shall be brazed in accordance with Section 1203.3.1.

1203.4 ABS plastic pipe. Joints between ABS plastic pipe or fittings shall be solvent-cemented or threaded joints conforming to Section 1203.3.

1203.5 Brass pipe. Joints between brass pipe or fittings shall be brazed, mechanical, threaded or welded joints conforming to Section 1203.3.

1203.6 Brass tubing. Joints between brass tubing or fittings shall be brazed, mechanical or soldered joints conforming to Section 1203.3.

1203.7 Copper or copper-alloy pipe. Joints between copper or copper-alloy pipe or fittings shall be brazed, mechanical, soldered, threaded or welded joints conforming to Section 1203.3.

1203.8 Copper or copper-alloy tubing. Joints between copper or copper-alloy tubing or fittings shall be brazed, mechanical or soldered joints conforming to Section 1203.3, flared joints conforming to Section 1203.8.1, push-fit joints conforming to Section 1203.8.2 or press-type joints conforming to Section 1203.8.3.

1203.8.1 Flared joints. Flared joints shall be made by a tool designed for that operation.

1203.8.2 Push-fit joints. Push-fit joints shall be installed in accordance with the manufacturer's instructions.

1203.8.3 Press joints. *Press joints* shall be installed in accordance with the manufacturer's instructions.

1203.9 CPVC plastic pipe. Joints between CPVC plastic pipe or fittings shall be solvent-cemented or threaded joints conforming to Section 1203.3.

1203.10 Polybutylene plastic pipe and tubing. Joints between polybutylene plastic pipe and tubing or fittings shall be mechanical joints conforming to Section 1203.3 or heat-fusion joints conforming to Section 1203.10.1.

1203.10.1 Heat-fusion joints. Joints shall be of the socket-fusion or butt-fusion type. Joint surfaces shall be clean and free of moisture. Joint surfaces shall be heated to melt temperatures and joined. The joint shall be undisturbed until cool. Joints shall be made in accordance with ASTM D 3309.

1203.11 Cross-linked polyethylene (PEX) plastic tubing. Joints between cross-linked polyethylene plastic tubing and fittings shall conform to Sections 1203.11.1 and 1203.11.2. Mechanical joints shall conform to Section 1203.3.

1203.11.1 Compression-type fittings. Where compression-type fittings include inserts and ferrules or O-rings, the fittings shall be installed without omitting the inserts and ferrules or O-rings.

1203.11.2 Plastic-to-metal connections. Soldering on the metal portion of the system shall be performed not less than 18 inches (457 mm) from a plastic-to-metal adapter in the same water line.

1203.12 PVC plastic pipe. Joints between PVC plastic pipe and fittings shall be solvent-cemented or threaded joints conforming to Section 1203.3.

1203.13 Steel pipe. Joints between steel pipe or fittings shall be mechanical joints that are made with an *approved* elastomeric seal, or shall be threaded or welded joints conforming to Section 1203.3.

1203.14 Steel tubing. Joints between steel tubing or fittings shall be mechanical or welded joints conforming to Section 1203.3.

1203.15 Polypropylene (PP) plastic. Joints between PP plastic pipe and fittings shall comply with Sections 1203.15.1 and 1203.15.2.

1203.15.1 Heat-fusion joints. Heat-fusion joints for polypropylene (PP) pipe and tubing joints shall be installed with socket-type heat-fused polypropylene fittings, electro-fusion polypropylene fittings or by butt fusion. Joint surfaces shall be clean and free from moisture. The joint shall be undisturbed until cool. Joints shall be made in accordance with ASTM F 2389.

1203.15.2 Mechanical and compression sleeve joints. Mechanical and compression sleeve joints shall be installed in accordance with the manufacturer's instructions.

1203.16 Raised temperature polyethylene (PE-RT) plastic tubing. Joints between raised temperature polyethylene tubing and fittings shall conform to Sections 1203.16.1 and

1203.16.2. Mechanical joints shall conform to Section 1203.3.

1203.16.1 Compression-type fittings. Where compression-type fittings include inserts and ferrules or O-rings, the fittings shall be installed without omitting the inserts and ferrules or O-rings.

1203.16.2 PE-RT-to-metal connections. Solder joints in a metal pipe shall not occur within 18 inches (457 mm) of a transition from such metal pipe to PE-RT pipe.

1203.17 Polyethylene/aluminum/polyethylene (PE-AL-PE) pressure pipe. Joints between polyethylene/aluminum/polyethylene pressure pipe and fittings shall conform to Sections 1203.17.1 and 1203.17.2. Mechanical joints shall comply with Section 1203.3.

1203.17.1 Compression-type fittings. Where compression-type fittings include inserts and ferrules or O-rings, the fittings shall be installed without omitting the inserts and ferrules or O-rings.

1203.17.2 PE-AL-PE-to-metal connections. Solder joints in a metal pipe shall not occur within 18 inches (457 mm) of a transition from such metal pipe to PE-AL-PE pipe.

1203.18 Cross-linked polyethylene/aluminum/cross-linked polyethylene (PEX-AL-PEX) pressure pipe. Joints between cross-linked polyethylene/aluminum/cross-linked polyethylene pressure pipe and fittings shall conform to Sections 1203.18.1 and 1203.18.2. Mechanical joints shall comply with Section 1203.3.

1203.18.1 Compression-type fittings. Where compression-type fittings include inserts and ferrules or O-rings, the fittings shall be installed without omitting the inserts and ferrules or O-rings.

1203.18.2 PEX-AL-PEX-to-metal connections. Solder joints in a metal pipe shall not occur within 18 inches (457 mm) of a transition from such metal pipe to PEX-AL-PEX pipe.

SECTION 1204
PIPE INSULATION

1204.1 Insulation characteristics. Pipe insulation installed in buildings shall conform to the requirements of the *International Energy Conservation Code*; shall be tested in accordance with ASTM E 84 or UL 723, using the specimen preparation and mounting procedures of ASTM E 2231; and shall have a maximum flame spread index of 25 and a smoke-developed index not exceeding 450. Insulation installed in an air *plenum* shall comply with Section 602.2.1.

Exception: The maximum flame spread index and smoke-developed index shall not apply to one- and two-family dwellings.

1204.2 Required thickness. Hydronic piping shall be insulated to the thickness required by the *International Energy Conservation Code*.

SECTION 1205
VALVES

1205.1 Where required. Shutoff valves shall be installed in hydronic piping systems in the locations indicated in Sections 1205.1.1 through 1205.1.6.

1205.1.1 Heat exchangers. Shutoff valves shall be installed on the supply and return side of a heat exchanger.

Exception: Shutoff valves shall not be required where heat exchangers are integral with a boiler; or are a component of a manufacturer's boiler and heat exchanger packaged unit and are capable of being isolated from the hydronic system by the supply and return valves required by Section 1005.1.

1205.1.2 Central systems. Shutoff valves shall be installed on the building supply and return of a central utility system.

1205.1.3 Pressure vessels. Shutoff valves shall be installed on the connection to any pressure vessel.

1205.1.4 Pressure-reducing valves. Shutoff valves shall be installed on both sides of a pressure-reducing valve.

1205.1.5 Equipment and appliances. Shutoff valves shall be installed on connections to mechanical *equipment* and appliances. This requirement does not apply to components of a hydronic system such as pumps, air separators, metering devices and similar *equipment*.

1205.1.6 Expansion tanks. Shutoff valves shall be installed at connections to nondiaphragm-type expansion tanks.

1205.2 Reduced pressure. A pressure relief valve shall be installed on the low-pressure side of a hydronic piping system that has been reduced in pressure. The relief valve shall be set at the maximum pressure of the system design. The valve shall be installed in accordance with Section 1006.

SECTION 1206
PIPING INSTALLATION

1206.1 General. Piping, valves, fittings and connections shall be installed in accordance with the conditions of approval.

1206.2 System drain down. Hydronic piping systems shall be designed and installed to permit the system to be drained. Where the system drains to the plumbing drainage system, the installation shall conform to the requirements of the *International Plumbing Code*.

Exception: The buried portions of systems embedded underground or under floors.

1206.3 Protection of potable water. The potable water system shall be protected from backflow in accordance with the *International Plumbing Code*.

1206.4 Pipe penetrations. Openings for pipe penetrations in walls, floors or ceilings shall be larger than the penetrating pipe. Openings through concrete or masonry building ele-ments shall be sleeved. The annular space surrounding pipe penetrations shall be protected in accordance with the *International Building Code*.

1206.5 Clearance to combustibles. A pipe in a hydronic piping system in which the exterior temperature exceeds 250°F (121°C) shall have a minimum *clearance* of 1 inch (25 mm) to combustible materials.

1206.6 Contact with building material. A hydronic piping system shall not be in direct contact with building materials that cause the piping material to degrade or corrode, or that interfere with the operation of the system.

1206.7 Water hammer. The flow velocity of the hydronic piping system shall be controlled to reduce the possibility of water hammer. Where a quick-closing valve creates water hammer, an *approved* water-hammer arrestor shall be installed. The arrestor shall be located within a range as specified by the manufacturer of the quick-closing valve.

1206.8 Steam piping pitch. Steam piping shall be installed to drain to the boiler or the steam trap. Steam systems shall not have drip pockets that reduce the capacity of the steam piping.

1206.9 Strains and stresses. Piping shall be installed so as to prevent detrimental strains and stresses in the pipe. Provisions shall be made to protect piping from damage resulting from expansion, contraction and structural settlement. Piping shall be installed so as to avoid structural stresses or strains within building components.

1206.9.1 Flood hazard. Piping located in a flood hazard area shall be capable of resisting hydrostatic and hydrodynamic loads and stresses, including the effects of buoyancy, during the occurrence of flooding to the *design flood elevation*.

1206.10 Pipe support. Pipe shall be supported in accordance with Section 305.

1206.11 Condensation. Provisions shall be made to prevent the formation of condensation on the exterior of piping.

SECTION 1207
TRANSFER FLUID

1207.1 Flash point. The flash point of transfer fluid in a hydronic piping system shall be not less than 50°F (28°C) above the maximum system operating temperature.

1207.2 Makeup water. The transfer fluid shall be compatible with the makeup water supplied to the system.

SECTION 1208
TESTS

1208.1 General. Hydronic piping systems shall be tested hydrostatically at one and one-half times the maximum system design pressure, but not less than 100 psi (689 kPa). The duration of each test shall be not less than 15 minutes.

SECTION 1209
EMBEDDED PIPING

1209.1 Materials. Piping for heating panels shall be standard-weight steel pipe, Type L copper tubing, polybutylene or other *approved* plastic pipe or tubing rated at 100 psi (689 kPa) at 180°F (82°C).

1209.2 Pressurizing during installation. Piping to be embedded in concrete shall be pressure tested prior to pouring concrete. During pouring, the pipe shall be maintained at the proposed operating pressure.

1209.3 Embedded joints. Joints of pipe or tubing that are embedded in a portion of the building, such as concrete or plaster, shall be in accordance with the requirements of Sections 1209.3.1 through 1209.3.4.

1209.3.1 Steel pipe joints. Steel pipe shall be welded by electrical arc or oxygen/acetylene method.

1209.3.2 Copper tubing joints. Copper tubing shall be joined by brazing complying with Section 1203.3.1.

1209.3.3 Polybutylene joints. Polybutylene pipe and tubing shall be installed in continuous lengths or shall be joined by heat fusion in accordance with Section 1203.10.1.

1209.3.4 Polyethylene of raised temperature (PE-RT) joints. PE-RT tubing shall be installed in continuous lengths or shall be joined by hydronic fittings listed in Table 1202.5.

1209.4 Not embedded related piping. Joints of other piping in cavities or running exposed shall be joined by *approved* methods in accordance with manufacturer's installation instructions and related sections of this code.

1209.5 Thermal barrier required. Radiant floor heating systems shall be provided with a thermal barrier in accordance with Sections 1209.5.1 through 1209.5.4.

> **Exception:** Insulation shall not be required in engineered systems where it can be demonstrated that the insulation will decrease the efficiency or have a negative effect on the installation.

1209.5.1 Slab-on-grade installation. Radiant piping utilized in slab-on-grade applications shall be provided with insulating materials installed beneath the piping having a minimum *R*-value of 5.

1209.5.2 Suspended floor installation. In suspended floor applications, insulation shall be installed in the joist bay cavity serving the heating space above and shall consist of materials having a minimum *R*-value of 11.

1209.5.3 Thermal break required. A thermal break shall be provided consisting of asphalt expansion joint materials or similar insulating materials at a point where a heated slab meets a foundation wall or other conductive slab.

1209.5.4 Thermal barrier material marking. Insulating materials utilized in thermal barriers shall be installed such that the manufacturer's *R*-value mark is readily observable upon inspection.

SECTION 1210
PLASTIC PIPE GROUND-SOURCE HEAT PUMP LOOP SYSTEMS

1210.1 Ground-source heat pump-loop water piping. Ground-source heat pump ground-loop piping and tubing material for water-based systems shall conform to the standards cited in this section.

1210.2 Used materials. Reused pipe, fittings, valves, and other materials shall not be permitted in ground-source heat pump loop systems.

1210.3 Material rating. Pipe and tubing shall be rated for the operating temperature and pressure of the ground-source heat pump loop system. Fittings shall be suitable for the pressure applications and recommended by the manufacturer for installation with the pipe and tubing material installed. Where used underground, materials shall be suitable for burial.

1210.4 Piping and tubing materials standards. Ground-source heat pump ground-loop pipe and tubing shall conform to the standards listed in Table 1210.4.

1210.5 Fittings. Ground-source heat pump pipe fittings shall be approved for installation with the piping materials to be installed, shall conform to the standards listed in Table 1210.5 and, if installed underground, shall be suitable for burial.

1210.6 Joints. Joints and connections shall be of an approved type. Joints and connections shall be tight for the pressure of the ground-source loop system. Joints used underground shall be approved for buried applications.

TABLE 1210.4
GROUND-SOURCE LOOP PIPE

MATERIAL	STANDARD (see Chapter 15)
Chlorinated polyvinyl chloride (CPVC)	ASTM D 2846; ASTM F 441; ASTM F 442
Cross-linked polyethylene (PEX)	ASTM F 876; ASTM F 877; CSA B137.5
Polyethylene/aluminum/polyethylene (PE-AL-PE) pressure pipe	ASTM F 1282; CSA B137.9
High-density polyethylene (HDPE)	ASTM D 2737; ASTM D 3035; ASTM F 714; AWWA C901; CSA B137.1; CSA C448; NSF 358-1
Polypropylene (PP-R)	ASTM F 2389; CSA B137.11
Polyvinyl chloride (PVC)	ASTM D 1785; ASTM D 2241
Raised temperature polyethylene (PE-RT)	ASTM F 2623

TABLE 1210.5
GROUND-SOURCE LOOP PIPE FITTINGS

PIPE MATERIAL	STANDARD (see Chapter 15)
Chlorinated polyvinyl chloride (CPVC)	ASTM D 2846; ASTM F 437; ASTM F 438; ASTM F 439; CSA B137.6
Cross-linked polyethylene (PEX)	ASTM F 877; ASTM F 1807; ASTM F 1960; ASTM F 2080; ASTM F 2159; ASTM F 2434; CSA B137.5
Polyethylene/aluminum/polyethylene (PE-AL-PE)	ASTM F 1282; ASTM F 2434; CSA B137.9
High Density Polyethylene (HDPE)	ASTM D 2683; ASTM D 3261; ASTM F 1055; CSA B137.1; CSA C448; NSF 358-1
Polypropylene (PP-R)	ASTM F 2389; CSA B137.11
Polyvinyl chloride (PVC)	ASTM D 2464; ASTM D 2466; ASTM D 2467; CSA B137.2; CSA B137.3
Raised temperature polyethylene (PE-RT)	ASTM D 3261; ASTM F 1807; ASTM F 2159; CSA B137.1

1210.6.1 Joints between different piping materials. Joints between different piping materials shall be made with approved transition fittings.

1210.6.2 Preparation of pipe ends. Pipe shall be cut square, be reamed, and be free of burrs and obstructions. CPVC, PE, and PVC pipe shall be chamfered. Pipe ends shall have full-bore openings and shall not be undercut.

1210.6.3 Joint preparation and installation. Where required by Sections 1210.6.4 through 1210.6.6, the preparation and installation of mechanical and thermoplastic-welded joints shall comply with Sections 1210.6.3.1 and 1210.6.3.2.

1210.6.3.1 Mechanical joints. Mechanical joints shall be installed in accordance with the manufacturer's instructions.

1210.6.3.2 Thermoplastic-welded joints. Joint surfaces for thermoplastic-welded joints shall be cleaned by an approved procedure. Joints shall be welded in accordance with the manufacturer's instructions.

1210.6.4 CPVC plastic pipe. Joints between CPVC plastic pipe or fittings shall be solvent-cemented or threaded joints complying with Section 1203.3.

1210.6.5 Cross-linked polyethylene (PEX) plastic tubing. Joints between cross-linked polyethylene plastic tubing and fittings shall comply with Sections 1210.6.5.1 and 1210.6.5.2. Mechanical joints shall comply with Section 1210.6.3.

1210.6.5.1 Compression-type fittings. Where compression-type fittings include inserts and ferrules or O-rings, the fittings shall be installed without omitting the inserts and ferrules or O-rings.

1210.6.5.2 Plastic-to-metal connections. Soldering on the metal portion of the system shall be performed not less than 18 inches (457 mm) from a plastic-to-metal adapter in the same water line.

1210.6.6 Polyethylene plastic pipe and tubing for ground-source heat pump loop systems. Joints between polyethylene plastic pipe and tubing or fittings for ground-source heat pump loop systems shall be heat fusion joints complying with Section 1210.6.6.1, electrofusion joints complying with Section 1210.6.6.2, or stab-type insertion joints complying with Section 1210.6.6.3.

1210.6.6.1 Heat-fusion joints. Joints shall be of the socket-fusion, saddle-fusion or butt-fusion type, joined in accordance with ASTM D 2657. Joint surfaces shall be clean and free from moisture. Joint surfaces shall be heated to melt temperatures and joined. The joint shall be undisturbed until cool. Fittings shall be manufactured in accordance with ASTM D 2683 or ASTM D 3261.

1210.6.6.2 Electrofusion joints. Joints shall be of the electrofusion type. Joint surfaces shall be clean and free from moisture, and scoured to expose virgin resin. Joint surfaces shall be heated to melt temperatures for the period of time specified by the manufacturer. The joint shall be undisturbed until cool. Fittings shall be manufactured in accordance with ASTM F 1055.

1210.6.6.3 Stab-type insert fittings. Joint surfaces shall be clean and free from moisture. Pipe ends shall be chamfered and inserted into the fittings to full depth. Fittings shall be manufactured in accordance with ASTM F 1924.

1210.6.7 Polypropylene (PP) plastic. Joints between PP plastic pipe and fittings shall comply with Sections 1210.6.7.1 and 1210.6.7.2.

1210.6.7.1 Heat-fusion joints. Heat-fusion joints for polypropylene (PP) pipe and tubing joints shall be installed with socket-type heat-fused polypropylene fittings, electrofusion polypropylene fittings or by butt fusion. Joint surfaces shall be clean and free from moisture. The joint shall be undisturbed until cool. Joints shall be made in accordance with ASTM F 2389.

1210.6.7.2 Mechanical and compression sleeve joints. Mechanical and compression sleeve joints shall be installed in accordance with the manufacturer's instructions.

1210.6.8 Raised temperature polyethylene (PE-RT) plastic tubing. Joints between raised temperature polyethylene tubing and fittings shall comply with Sections 1210.6.8.1 and 1210.6.8.2. Mechanical joints shall comply with Section 1210.6.3.

1210.6.8.1 Compression-type fittings. Where compression-type fittings include inserts and ferrules or O-rings, the fittings shall be installed without omitting the inserts and ferrules or O-rings.

1210.6.8.2 PE-RT-to-metal connections. Solder joints in a metal pipe shall not occur within 18 inches (457 mm) of a transition from such metal pipe to PE-RT pipe.

1210.6.9 PVC plastic pipe. Joints between PVC plastic pipe and fittings shall be solvent-cemented or threaded joints comply with Section 1203.3.

1210.7 Shutoff valves. Shutoff valves shall be installed in ground-source loop piping systems in the locations indicated in Sections 1210.7.1 through 1210.7.7.

1210.7.1 Heat exchangers. Shutoff valves shall be installed on the supply and return side of a heat exchanger.

Exception: Shutoff valves shall not be required where heat exchangers are integral with a boiler or are a component of a manufacturer's boiler and heat exchanger packaged unit and are capable of being isolated from the hydronic system by the supply and return valves required by Section 1005.1.

1210.7.2 Central systems. Shutoff valves shall be installed on the building supply and return of a central utility system.

1210.7.3 Pressure vessels. Shutoff valves shall be installed on the connection to any pressure vessel.

1210.7.4 Pressure-reducing valves. Shutoff valves shall be installed on both sides of a pressure-reducing valve.

1210.7.5 Equipment and appliances. Shutoff valves shall be installed on connections to mechanical *equipment* and appliances. This requirement does not apply to components of a ground-source loop system such as pumps, air separators, metering devices, and similar *equipment*.

1210.7.6 Expansion tanks. Shutoff valves shall be installed at connections to nondiaphragm-type expansion tanks.

1210.7.7 Reduced pressure. A pressure relief valve shall be installed on the low-pressure side of a hydronic piping system that has been reduced in pressure. The relief valve shall be set at the maximum pressure of the system design. The valve shall be installed in accordance with Section 1006.

1210.8 Installation. Piping, valves, fittings, and connections shall be installed in accordance with the conditions of approval.

1210.8.1 Protection of potable water. Where ground-source heat pump ground-loop systems have a connection to a potable water supply, the potable water system shall be protected from backflow in accordance with the *International Plumbing Code*.

1210.8.2 Pipe penetrations. Openings for pipe penetrations in walls, floors and ceilings shall be larger than the penetrating pipe. Openings through concrete or masonry building elements shall be sleeved. The annular space surrounding pipe penetrations shall be protected in accordance with the *International Building Code*.

1210.8.3 Clearance from combustibles. A pipe in a ground-source heat pump piping system having an exterior surface temperature exceeding 250°F (121°C) shall have a minimum *clearance* of 1 inch (25 mm) from combustible materials.

1210.8.4 Contact with building material. A ground-source heat pump ground-loop piping system shall not be in direct contact with building materials that cause the piping or fitting material to degrade or corrode, or that interfere with the operation of the system.

1210.8.5 Strains and stresses. Piping shall be installed so as to prevent detrimental strains and stresses in the pipe. Provisions shall be made to protect piping from damage resulting from expansion, contraction and structural settlement. Piping shall be installed so as to avoid structural stresses or strains within building components.

1210.8.6 Flood hazard. Piping located in a flood hazard area shall be capable of resisting hydrostatic and hydrodynamic loads and stresses, including the effects of buoyancy, during the occurrence of flooding to the *design flood elevation*.

1210.8.7 Pipe support. Pipe shall be supported in accordance with Section 305.

1210.8.8 Velocities. Ground-source heat pump ground-loop systems shall be designed so that the flow velocities do not exceed the maximum flow velocity recommended by the pipe and fittings manufacturer and shall be controlled to reduce the possibility of water hammer.

1210.8.9 Labeling and marking. Ground-source heat pump ground-loop system piping shall be marked with tape, metal tags or other method where it enters a building indicating "GROUND-SOURCE HEAT PUMP LOOP SYSTEM." The marking shall indicate any antifreeze used in the system by name and concentration.

1210.8.10 Chemical compatibility. Antifreeze and other materials used in the system shall be chemically compatible with the pipe, tubing, fittings, and mechanical systems.

1210.9 Makeup water. The transfer fluid shall be compatible with the makeup water supplied to the system.

1210.10 Tests. Before connection header trenches are backfilled, the assembled loop system shall be pressure tested with water at 100 psi (689 kPa) for 15 minutes, in which time there shall not be observed leaks. Flow and pressure loss testing shall be performed and the actual flow rates and pressure drops shall be compared to the calculated design values. If actual flow rate or pressure drop values differ from calculated design values by more than 10 percent, the cause shall be identified and corrective action taken.

1210.11 Embedded piping. Ground-source heat pump ground-loop piping to be embedded in concrete shall be pressure tested prior to pouring concrete. During pouring, the pipe shall be maintained at the proposed operating pressure.

CHAPTER 13

FUEL OIL PIPING AND STORAGE

SECTION 1301
GENERAL

1301.1 Scope. This chapter shall govern the design, installation, construction and repair of fuel-oil storage and piping systems. The storage of fuel oil and flammable and combustible liquids shall be in accordance with Chapters 6 and 57 of the *International Fire Code.*

1301.2 Storage and piping systems. Fuel-oil storage systems shall comply with Section 603.3 of the *International Fire Code.* Fuel-oil piping systems shall comply with the requirements of this code.

1301.3 Fuel type. An *appliance* shall be designed for use with the type of fuel to which it will be connected. Such *appliance* shall not be converted from the fuel specified on the rating plate for use with a different fuel without securing reapproval from the code official.

1301.4 Fuel tanks, piping and valves. The tank, piping and valves for appliances burning oil shall be installed in accordance with the requirements of this chapter. Where an oil burner is served by a tank, any part of which is above the level of the burner inlet connection and where the fuel supply line is taken from the top of the tank, an *approved* antisiphon valve or other siphon-breaking device shall be installed in lieu of the shutoff valve.

1301.5 Tanks abandoned or removed. All exterior above-grade fill piping shall be removed when tanks are abandoned or removed. Tank abandonment and removal shall be in accordance with Section 5704.2.13 of the *International Fire Code.*

SECTION 1302
MATERIAL

1302.1 General. Piping materials shall conform to the standards cited in this section.

1302.2 Rated for system. All materials shall be rated for the operating temperatures and pressures of the system, and shall be compatible with the type of liquid.

1302.3 Pipe standards. Fuel oil pipe shall comply with one of the standards listed in Table 1302.3.

1302.4 Nonmetallic pipe. Nonmetallic pipe shall be *listed* and *labeled* as being acceptable for the intended application for flammable and combustible liquids. Nonmetallic pipe shall be installed only outside, underground.

1302.5 Fittings and valves. Fittings and valves shall be *approved* for the piping systems, and shall be compatible with, or shall be of the same material as, the pipe or tubing.

1302.6 Bending of pipe. Pipe shall be *approved* for bending. Pipe bends shall be made with *approved equipment.* The bend shall not exceed the structural limitations of the pipe.

TABLE 1302.3
FUEL OIL PIPING

MATERIAL	STANDARD (see Chapter 15)
Copper or copper-alloy pipe	ASTM B 42; ASTM B 43; ASTM B 302
Copper or copper-alloy tubing (Type K, L or M)	ASTM B 75; ASTM B 88; ASTM B 280; ASME B 16.51
Labeled pipe	(See Section 1302.4)
Nonmetallic pipe	ASTM D 2996
Steel pipe	ASTM A 53; ASTM A 106
Steel tubing	ASTM A 254; ASTM A 539

1302.7 Pumps. Pumps that are not part of an *appliance* shall be of a positive-displacement type. The pump shall automatically shut off the supply when not in operation. Pumps shall be *listed* and *labeled* in accordance with UL 343.

1302.8 Flexible connectors and hoses. Flexible connectors and hoses shall be *listed* and *labeled* in accordance with UL 536.

SECTION 1303
JOINTS AND CONNECTIONS

1303.1 Approval. Joints and connections shall be *approved* and of a type *approved* for fuel-oil piping systems. Threaded joints and connections shall be made tight with suitable lubricant or pipe compound. Unions requiring gaskets or packings, right or left couplings, and sweat fittings employing solder having a melting point of less than 1,000°F (538°C) shall not be used in oil lines. Cast-iron fittings shall not be used. Joints and connections shall be tight for the pressure required by test.

1303.1.1 Joints between different piping materials. Joints between different piping materials shall be made with *approved* adapter fittings. Joints between different metallic piping materials shall be made with *approved* dielectric fittings or brass converter fittings.

1303.2 Preparation of pipe ends. Pipe shall be cut square, reamed and chamfered and be free from all burrs and obstructions. Pipe ends shall have full-bore openings and shall not be undercut.

1303.3 Joint preparation and installation. Where required by Sections 1303.4 through 1303.10, the preparation and installation of brazed, mechanical, threaded and welded joints shall comply with Sections 1303.3.1 through 1303.3.4.

1303.3.1 Brazed joints. All joint surfaces shall be cleaned. An *approved* flux shall be applied where required. The joints shall be brazed with a filler metal conforming to AWS A5.8.

1303.3.2 Mechanical joints. Mechanical joints shall be installed in accordance with the manufacturer's instructions. Press connect joints shall conform to one of the standards listed in Table 1302.3.

1303.3.3 Threaded joints. Threads shall conform to ASME B1.20.1. Pipe-joint compound or tape shall be applied on the male threads only.

1303.3.4 Welded joints. All joint surfaces shall be cleaned by an *approved* procedure. The joint shall be welded with an *approved* filler metal.

1303.4 Brass pipe. Joints between brass pipe or fittings shall be brazed, mechanical, threaded or welded joints complying with Section 1303.3.

1303.5 Brass tubing. Joints between brass tubing or fittings shall be brazed or mechanical joints complying with Section 1303.3.

1303.6 Copper or copper-alloy pipe. Joints between copper or copper-alloy pipe or fittings shall be brazed, mechanical, threaded or welded joints complying with Section 1303.3.

1303.7 Copper or copper-alloy tubing. Joints between copper or copper-alloy tubing or fittings shall be brazed, mechanical joints complying with Section 1303.3, press connect joints that conform to one of the standards in Table 1302.3 or flared joints. Flared joints shall be made by a tool designed for that operation.

1303.8 Nonmetallic pipe. Joints between nonmetallic pipe or fittings shall be installed in accordance with the manufacturer's instructions for the *labeled* pipe and fittings.

1303.9 Steel pipe. Joints between steel pipe or fittings shall be threaded or welded joints complying with Section 1303.3 or mechanical joints complying with Section 1303.9.1.

1303.9.1 Mechanical joints. Joints shall be made with an *approved* elastomeric seal. Mechanical joints shall be installed in accordance with the manufacturer's instructions. Mechanical joints shall be installed outside, underground, unless otherwise *approved*.

1303.10 Steel tubing. Joints between steel tubing or fittings shall be mechanical or welded joints complying with Section 1303.3.

1303.11 Piping protection. Proper allowance shall be made for expansion, contraction, jarring and vibration. Piping other than tubing, connected to underground tanks, except straight fill lines and test wells, shall be provided with flexible connectors, or otherwise arranged to permit the tanks to settle without impairing the tightness of the piping connections.

SECTION 1304
PIPING SUPPORT

1304.1 General. Pipe supports shall be in accordance with Section 305.

SECTION 1305
FUEL OIL SYSTEM INSTALLATION

1305.1 Size. The fuel oil system shall be sized for the maximum capacity of fuel oil required. The minimum size of a supply line shall be $^3/_8$-inch (9.5 mm) inside diameter nominal

pipe or $^3/_8$-inch (9.5 mm) od tubing. The minimum size of a return line shall be $^1/_4$-inch (6.4 mm) inside diameter nominal pipe or $^5/_{16}$-inch (7.9 mm) outside diameter tubing. Copper tubing shall have 0.035-inch (0.9 mm) nominal and 0.032-inch (0.8 mm) minimum wall thickness.

1305.2 Protection of pipe, equipment and appliances. Fuel oil pipe, *equipment* and appliances shall be protected from physical damage.

1305.2.1 Flood hazard. Fuel oil pipe, equipment and appliances located in flood hazard areas shall be located above the elevation required by Section 1612 of the *International Building Code* for utilities and attendant equipment or shall be capable of resisting hydrostatic and hydrodynamic loads and stresses, including the effects of buoyancy, during the occurrence of flooding up to such elevation.

1305.3 Supply piping. Supply piping shall connect to the top of the fuel oil tank. Fuel oil shall be supplied by a transfer pump or automatic pump or by other *approved* means.

Exception: This section shall not apply to inside or above-ground fuel oil tanks.

1305.4 Return piping. Return piping shall connect to the top of the fuel oil tank. Valves shall not be installed on return piping.

1305.5 System pressure. The system shall be designed for the maximum pressure required by the fuel-oil-burning *appliance*. Air or other gases shall not be used to pressurize tanks.

1305.6 Fill piping. A fill pipe shall terminate outside of a building at a point not less than 2 feet (610 mm) from any building opening at the same or lower level. A fill pipe shall terminate in a manner designed to minimize spilling when the filling hose is disconnected. Fill opening shall be equipped with a tight metal cover designed to discourage tampering.

1305.7 Vent piping. Liquid fuel vent pipes shall terminate outside of buildings at a point not less than 2 feet (610 mm) measured vertically or horizontally from any building opening. Outer ends of vent pipes shall terminate in a weatherproof vent cap or fitting or be provided with a weatherproof hood. Vent caps shall have a minimum free open area equal to the cross-sectional area of the vent pipe and shall not employ screens finer than No. 4 mesh. Vent pipes shall terminate sufficiently above the ground to avoid being obstructed with snow or ice. Vent pipes from tanks containing heaters shall be extended to a location where oil vapors discharging from the vent will be readily diffused. If the static head with a vent pipe filled with oil exceeds 10 pounds per square inch (psi) (69 kPa), the tank shall be designed for the maximum static head that will be imposed.

Liquid fuel vent pipes shall not be cross connected with fill pipes, lines from burners or overflow lines from auxiliary tanks.

SECTION 1306
OIL GAUGING

1306.1 Level indication. Tanks in which a constant oil level is not maintained by an automatic pump shall be equipped with a method of determining the oil level.

1306.2 Test wells. Test wells shall not be installed inside buildings. For outside service, test wells shall be equipped with a tight metal cover designed to discourage tampering.

1306.3 Inside tanks. The gauging of inside tanks by means of measuring sticks shall not be permitted. An inside tank provided with fill and vent pipes shall be provided with a device to indicate either visually or audibly at the fill point when the oil in the tank has reached a predetermined safe level.

1306.4 Gauging devices. Gauging devices such as liquid level indicators or signals shall be designed and installed so that oil vapor will not be discharged into a building from the liquid fuel supply system. Liquid-level indicating gauges shall comply with UL 180.

1306.5 Gauge glass. A tank used in connection with any oil burner shall not be equipped with a glass gauge or any gauge which, when broken, will permit the escape of oil from the tank.

SECTION 1307
FUEL OIL VALVES

1307.1 Building shutoff. A shutoff valve shall be installed on the fuel-oil supply line at the entrance to the building. Inside or above-ground tanks are permitted to have valves installed at the tank. The valve shall be capable of stopping the flow of fuel oil to the building or to the *appliance* served where the valve is installed at a tank inside the building. Valves shall comply with UL 842.

1307.2 Appliance shutoff. A shutoff valve shall be installed at the connection to each *appliance* where more than one fuel-oil-burning *appliance* is installed.

1307.3 Pump relief valve. A relief valve shall be installed on the pump discharge line where a valve is located downstream of the pump and the pump is capable of exceeding the pressure limitations of the fuel oil system.

1307.4 Fuel-oil heater relief valve. A relief valve shall be installed on the discharge line of fuel-oil-heating appliances.

1307.5 Relief valve operation. The relief valve shall discharge fuel oil when the pressure exceeds the limitations of the system. The discharge line shall connect to the fuel oil tank.

SECTION 1308
TESTING

1308.1 Testing required. Fuel oil piping shall be tested in accordance with NFPA 31.

CHAPTER 14

SOLAR SYSTEMS

SECTION 1401
GENERAL

1401.1 Scope. This chapter shall govern the design, construction, installation, *alteration* and repair of systems, *equipment* and appliances intended to utilize solar energy for space heating or cooling, domestic hot water heating, swimming pool heating or process heating.

1401.2 Potable water supply. Potable water supplies to solar systems shall be protected against contamination in accordance with the *International Plumbing Code*.

> **Exception:** Where all solar system piping is a part of the potable water distribution system, in accordance with the requirements of the *International Plumbing Code*, and all components of the piping system are *listed* for potable water use, cross-connection protection measures shall not be required.

1401.3 Heat exchangers. Heat exchangers used in domestic water-heating systems shall be *approved* for the intended use. The system shall have adequate protection to ensure that the potability of the water supply and distribution system is properly safeguarded.

1401.4 Solar energy equipment and appliances. Solar energy *equipment* and appliances shall conform to the requirements of this chapter and shall be installed in accordance with the manufacturer's instructions.

1401.5 Ducts. Ducts utilized in solar heating and cooling systems shall be constructed and installed in accordance with Chapter 6 of this code.

SECTION 1402
INSTALLATION

1402.1 Access. Access shall be provided to solar energy *equipment* and appliances for maintenance. Solar systems and appurtenances shall not obstruct or interfere with the operation of any doors, windows or other building components requiring operation or access.

1402.2 Protection of equipment. Solar *equipment* exposed to vehicular traffic shall be installed not less than 6 feet (1829 mm) above the finished floor.

> **Exception:** This section shall not apply where the *equipment* is protected from motor vehicle impact.

1402.3 Controlling condensation. Where attics or structural spaces are part of a passive solar system, ventilation of such spaces, as required by Section 406, is not required where other *approved* means of controlling condensation are provided.

1402.4 Roof-mounted collectors. Roof-mounted solar collectors that also serve as a roof covering shall conform to the requirements for roof coverings in accordance with the *International Building Code*.

> **Exception:** The use of plastic solar collector covers shall be limited to those *approved* plastics meeting the requirements for plastic roof panels in the *International Building Code*.

1402.4.1 Collectors mounted above the roof. Where mounted on or above the roof covering, the collector array and supporting construction shall be constructed of noncombustible materials or fire-retardant-treated wood conforming to the *International Building Code* to the extent required for the type of roof construction of the building to which the collectors are accessory.

> **Exception:** The use of plastic solar collector covers shall be limited to those *approved* plastics meeting the requirements for plastic roof panels in the *International Building Code*.

1402.5 Equipment. The solar energy system shall be equipped in accordance with the requirements of Sections 1402.5.1 through 1402.5.4.

1402.5.1 Pressure and temperature. Solar energy system components containing pressurized fluids shall be protected against pressures and temperatures exceeding design limitations with a pressure and temperature relief valve. Each section of the system in which excessive pressures are capable of developing shall have a relief device located so that a section cannot be valved off or otherwise isolated from a relief device. Relief valves shall comply with the requirements of Section 1006.4 and discharge in accordance with Section 1006.6.

1402.5.2 Vacuum. The solar energy system components that are subjected to a vacuum while in operation or during shutdown shall be designed to withstand such vacuum or shall be protected with vacuum relief valves.

1402.5.3 Protection from freezing. System components shall be protected from damage by freezing of heat transfer liquids at the lowest ambient temperatures that will be encountered during the operation of the system.

1402.5.4 Expansion tanks. Liquid single-phase solar energy systems shall be equipped with expansion tanks sized in accordance with Section 1009.

1402.6 Penetrations. Roof and wall penetrations shall be flashed and sealed to prevent entry of water, rodents and insects.

1402.7 Filtering. Air transported to occupied spaces through rock or dust-producing materials by means other than natural convection shall be filtered at the outlet from the heat storage system.

SECTION 1403
HEAT TRANSFER FLUIDS

1403.1 Flash point. The flash point of the actual heat transfer fluid utilized in a solar system shall be not less than 50°F (28°C) above the design maximum nonoperating (no-flow) temperature of the fluid attained in the collector.

1403.2 Flammable gases and liquids. A flammable liquid or gas shall not be utilized as a heat transfer fluid. The flash point of liquids used in occupancies classified in Group H or F shall not be lower unless *approved*.

SECTION 1404
MATERIALS

1404.1 Collectors. Factory-built collectors shall be *listed* and *labeled*, and bear a label showing the manufacturer's name and address, model number, collector dry weight, collector maximum allowable operating and nonoperating temperatures and pressures, minimum allowable temperatures and the types of heat transfer fluids that are compatible with the collector. The label shall clarify that these specifications apply only to the collector.

1404.2 Thermal storage units. Pressurized thermal storage units shall be *listed* and *labeled*, and bear a label showing the manufacturer's name and address, model number, serial number, storage unit maximum and minimum allowable operating temperatures, storage unit maximum and minimum allowable operating pressures and the types of heat transfer fluids compatible with the storage unit. The label shall clarify that these specifications apply only to the thermal storage unit.

CHAPTER 15

REFERENCED STANDARDS

This chapter lists the standards that are referenced in various sections of this document. The standards are listed herein by the promulgating agency of the standard, the standard identification, the effective date and title, and the section or sections of this document that reference the standard. The application of the referenced standards shall be as specified in Section 102.8.

ACCA

Air Conditioning Contractors of America
2800 Shirlington Road, Suite 300
Arlington, VA 22206

Standard reference number	Title	Referenced in code section number
Manual D—2011	Residential Duct Systems. 601.4, 603.2	
183—2007 (reaffirmed 2011)	Peak Cooling and Heating Load Calculations in Buildings Except Low-rise Residential Buildings 312.1	

AHRI

Air-Conditioning, Heating and Refrigeration Institute
4100 North Fairfax Drive, Suite 200
Arlington, VA 22203

Standard reference number	Title	Referenced in code section number
700—2011 with Addendum 1	Purity Specifications for Fluorocarbon and Other Refrigerants 1102.2.2.3	

AMCA

Air Movement and Control Association International
30 West University Drive
Arlington Heights, IL 60004

Standard reference number	Title	Referenced in code section number
550—08	Test Method for High Velocity Wind Driven Rain Resistant Louvers 401.5, 501.3.2	

ANSI

American National Standards Institute
11 West 42nd Street
New York, NY 10036

Standard reference number	Title	Referenced in code section number
Z21.8—1994 (R2002)	Installation of Domestic Gas Conversion Burners. 919.1	

ASHRAE

ASHRAE
1791 Tullie Circle, NE
Atlanta, GA 30329

Standard reference number	Title	Referenced in code section number
ASHRAE—2013	ASHRAE Fundamentals Handbook	603.2
15—2013	Safety Standard for Refrigeration Systems	1101.6, 1105.8, 1108.1
34—2013	Designation and Safety Classification of Refrigerants	202, 1102.2.1, 1103.1
62.1—2013	Ventilation for Acceptable Indoor Air Quality	403.3.1.1.2.3.2
170—2008	Ventilation of Health Care Facilities	407
180—2012	Standard Practice for Inspection and Maintenance of Commercial Building HVAC Systems	102.3

ASME

American Society of Mechanical Engineers
Three Park Avenue
New York, NY 10016-5990

Standard reference number	Title	Referenced in code section number
B1.20.1—1983 (R2006)	Pipe Threads, General Purpose (Inch)	1203.3.5, 1303.3.3
B16.3—2011	Malleable Iron Threaded Fittings, Classes 150 & 300	Table 1202.5
B16.5—2009	Pipe Flanges and Flanged Fittings NPS $^1/_2$ through NPS 24	Table 1202.5
B16.9—2007	Factory Made Wrought Steel Buttwelding Fittings	Table 1202.5
B16.11—2011	Forged Fittings, Socket-welding and Threaded	Table 1202.5
B16.15—2011	Cast Bronze Threaded Fittings	Table 1202.5
B16.18—2012	Cast Copper Alloy Solder Joint Pressure Fittings	513.13.1, Table 1202.5
B16.22—(R2010)	Wrought Copper and Copper Alloy Solder Joint Pressure Fittings	513.13.1, Table 1202.5
B16.24—2011	Cast Copper Alloy Pipe Flanges and Flanged Fittings: Class 150, 300, 400, 600, 900, 1500 and 2500	Table 1202.5
B16.26—2011	Cast Copper Alloy Fittings for Flared Copper Tubes	Table 1202.5
B16.28—1994	Wrought Steel Buttwelding Short Radius Elbows and Returns	Table 1202.5
B16.51—2011	Copper and Copper Alloy Press-Connect Pressure Fittings	Table 1202.5
B31.5—2010	Refrigeration Piping and Heat Transfer Components	1107.1
B31.9—2011	Building Services Piping	1201.3
BPVC—2010/ 2011 addenda	ASME Boiler & Pressure Vessel Code–07 Edition	1009.2, 1003.1, 1004.1, 1011.1
CSD-1—2011	Controls and Safety Devices for Automatically Fired Boilers	1004.1

ASSE

American Society of Safety Engineers
1800 East Oakton Street
Des Plaines, IL 60018

Standard reference number	Title	Referenced in code section number
ANSI/ASSE Z359.1-2007	Safety Requirements for Personal Fall Arrest Systems, Subsystems and Components, Part of the Fall Protection Code	304.11

ASSE

American Society of Sanitary Engineering
901 Canterbury, Suite A
Westlake, OH 44145

Standard reference number	Title	Referenced in code section number
1017—2010	Performance Requirements for Temperature Actuated Mixing Values for Hot Water Distribution Systems	1002.2.2

ASTM

ASTM International
100 Barr Harbor Drive
West Conshohocken, PA 19428

Standard reference number	Title	Referenced in code section number
A 53/A 53M—12	Specification for Pipe, Steel, Black and Hot-dipped, Zinc-coated Welded and Seamless	Table 1202.4, Table 1202.5, Table 1302.3
A 106/A106M—11	Specification for Seamless Carbon Steel Pipe for High-Temperature Service	Table 1302.3, Table 1202.5 Table 1202.4 Table 1302.3
A 126—09	Specification for Gray Iron Castings for Valves, Flanges and Pipe Fittings	Table 1202.5
A 234/A234M—11a	Standard Specification for Piping Fittings of Wrought Carbon Steel And Alloy Steel For Moderate and High Temperature Service	Table 1202.5
A 254—97 (2007)	Specification for Copper Brazed Steel Tubing	Table 1202.4, Table 1302.3
A 395/A395M-99(2009)	Standard Specification for Ferritic Ductile Iron Pressure-Retaining Castings for Use at Elevated Temperatures	Table 1202.5, Table 1302.3
A 420/A 420M—10A	Specification for Piping Fittings of Wrought Carbon Steel and Alloy Steel for Low-Temperature Service	Table 1202.5
A 536-84(2009)	Standard Specification for Ductile Iron Castings	Table 1202.5
A 539—99	Specification for Electric-resistance-welded Coiled Steel Tubing for Gas and Fuel Oil Lines	Table 1302.3
B 32—08	Specification for Solder Metal	1203.3.3
B 42—10	Specification for Seamless Copper Pipe, Standard Sizes	513.13.1, 1107.5.2, Table 1302.3 Table 1202.4, Table 1302.3
B 43—09	Specification for Seamless Red Brass Pipe, Standard Sizes	513.13.1, 1107.5.2, Table 1302.3 Table 1202.4, Table 1302.3
B 68—11	Specification for Seamless Copper Tube, Bright Annealed	513.13.1
B75—11	Specification for Seamless Copper Tube	Table 1302.3
B 88—09	Specification for Seamless Copper Water Tube	513.13.1, 1107.5.3, Table 1202.4, Table 1302.3
B 135—10	Specification for Seamless Brass Tube	Table 1202.4
B 251—10	Specification for General Requirements for Wrought Seamless Copper and Copper-alloy Tube	513.13.1, Table 1202.4
B 280—08	Specification for Seamless Copper Tube for Air Conditioning and Refrigeration Field Service	513.13.1, 1107.5.3, Table 1302.3
B 302—12	Specification for Threadless Copper Pipe, Standard Sizes	Table 1202.4, Table 1302.3
B 813—10	Specification for Liquid and Paste Fluxes for Soldering of Copper and Copper Alloy Tube	1203.3.3
C 315—07(2011)	Specification for Clay Flue Liners and Chimney Pots	801.16.1, Table 803.10.4
C 411—11	Test Method for Hot-surface Performance of High-temperature Thermal Insulation	604.3
D 56—05(2010)	Test Method for Flash Point by Tag Closed Tester	202
D 93—12	Test Method for Flash Point of Pensky-Martens Closed Cup Tester	202
D 1527—99(2005)	Specification for Acrylonitrile-Butadiene-Styrene (ABS) Plastic Pipe, Schedules 40 and 80	Table 1202.4
D 1693—13	Test Method for Environmental Stress-Cracking of Ethylene Plastics	Table 1202.4
D 1785—12	Specification for Poly (Vinyl Chloride)(PVC) Plastic Pipe, Schedules 40, 80 and 120	Table 1202.4, Table 1210.4
D 2235—04(2011)	Specifications for Solvent Cement for Acrylonitrile-Butadiene-Styrene (ABS) Plastic Pipe and Fittings	1203.3.4
D 2241—09	Specification for Poly (Vinyl Chloride)(PVC) Pressure-rated Pipe (SDR-Series)	Table 1202.4, Table 1210.4
D 2282—99(2005)	Specification for Acrylonitrile-Butadiene-Styrene (ABS) Plastic Pipe (SDR-PR)	Table 1202.4
D 2412—11	Test Method for Determination of External Loading Characteristics of Plastic Pipe by Parallel-plate Loading	603.8.3
D2464—13	Standard Specification for Threaded Poly(Vinyl Chloride) (PVC) Plastic Pipe Fittings, Schedule 80	Table 1210.5
D 2466—06	Specification for Poly (Vinyl Chloride)(PVC) Plastic Pipe Fittings, Schedule 40	Table 1202.5, Table 1210.5
D 2467—06	Specification for Poly (Vinyl Chloride)(PVC) Plastic Pipe Fittings, Schedule 80	Table 1202.5, Table 1210.5
D 2564—12	Specification for Solvent Cements for Poly (Vinyl Chloride) (PVC) Plastic Piping Systems	1203.3.4
D 2657—07	Standard Practice for Heat Fusion Joining of Polyolefin Pipe and Fittings	Table 1210.5
D 2683—2010E1	Specification for Socket-type Polyethylene Fittings for Outside Diameter-controlled Polyethylene Pipe and Tubing	Table 1210.5, 1210.6.6.1
D2737—12A	Standard Specification for Polyethylene (PE) Plastic Tubing	Table 1210.4
D 2846/D 2846M—09BE1	Specification for Chlorinated Poly (Vinyl Chloride) (CPVC) Plastic Hot and Cold Water Distribution Systems	Table 1210.4, Table 1202.4, 1203.3.4
D 2996—01(2007)e01	Specification for Filament-wound Fiberglass (Glass Fiber Reinforced Thermosetting Resin) Pipe	Table 1302.3

ASTM—continued

D 3035—2012E1	Specification for Polyethylene (PE) Plastic Pipe (DR-PR) Based on Controlled Outside Diameter	Table 1210.4
D 3261—12	Specification for Butt Heat Fusion Polyethylene (PE) Plastic Fittings for Polyethylene (PE) Plastic Pipe and Tubing	Table 1210.5, 1210.6.6.1
D 3278—1996(2011)	Test Methods for Flash Point of Liquids by Small Scale Closed-cup Apparatus	202
D 3309—96a(2002)	Specification for Polybutylene (PB) Plastic Hot and Cold Water Distribution Systems	Table 1202.4, 1203.10.1
E 84—2013A	Test Method for Surface Burning Characteristics of Building Materials	202, 510.9, 602.2, 602.2.1, 602.2.1.6, 602.2.1.6.1, 602.2.1.6.2, 602.2.1.6.3, 602.2.1.7, 604.3, 1204.1
E 119—2012a	Test Method for Fire Tests of Building Construction and Materials	607.5.2, 607.5.5, 607.6.1, 607.2.1
E 136—2012	Test Method for Behavior of Materials in a Vertical Tube Furnace at 750 Degrees C	202
E 814—2013	Test Method for Fire Tests of Through-penetration Fire Stops	506.3.11.2, 506.3.11.3
E 1509—12	Specification for Room Heaters, Pellet Fuel-burning Type	904.1
E 2231—09	Standard Practice For Specimen Preparation and Mounting of Pipe and Duct Insulation Materials to Assess Surface Burning Characteristics	604.3, 1204.1
E 2336—04(2013)	Standard Test Methods for Fire Resistive Grease Duct Enclosure Systems	506.3.6, 506.3.11.2
F437—09	Specification for Threaded Chlorinated Poly (Vinyl Chloride) (CPVC) Plastic Pipe Fittings, Schedule 80	Table 1210.5
F 438—09	Specification for Socket Type Chlorinated Poly (Vinyl Chloride) (CPVC) Plastic Pipe Fittings, Schedule 40	Table 1202.5, Table 1210.5
F 439—12	Specification for Socket Type Chlorinated Poly (Vinyl Chloride) (CPVC) Plastic Pipe Fittings, Schedule 80	Table 1202.5, Table 1210.5
F 441/F 441M—13	Specification for Chlorinated Poly (Vinyl Chloride) (CPVC) Plastic Pipe, Schedules 40 and 80	Table 1202.4, Table 1210.4
F 442/F 442M—13	Specification for Chlorinated Poly (Vinyl Chloride) (CPVC) Plastic Pipe (SDR-PR)	Table 1202.4, Table 1210.4
F 493—10	Specification for Solvent Cements for Chlorinated Poly (Vinyl Chloride) (CPVC) Plastic Pipe and Fittings	1203.3.4
F 714—13	Standard Specification for Polyethylene (PE) Plastic Pipe (SDR-PR) based on Outside Diameter	Table 1210.4
F 876—13	Specification for Cross-linked Polyethylene (PEX) Tubing	Table 1202.4, Table 1210.4
F 877—2011A	Specification for Cross-linked Polyethylene (PEX) Plastic Hot and Cold Water Distribution Systems	Table 1202.4, Table 1202.5, Table 1210.4
F 1055—13	Specification for Electrofusion Type Polyethylene Fittings for Outside Diameter Controlled Polyethylene and Cross-linked Polyethylene Pipe and Tubing	Table 1210.5, 1210.6.6.2
F 1281—11	Specification for Cross-linked Polyethylene/Aluminum/Crosslinked Polyethylene (PEX-AL-PEX) Pressure Pipe	Table 1202.4
F 1282—10	Standard Specification for Polyethylene/Aluminum/ Polyethylene (PE-AL-PE) Composite Pressure Pipe	Table 1202.4. Table 1210.4, Table 1210.5
F 1476—07	Specification for Performance of Gasketed Mechanical Couplings for Use in Piping Applications	Table 1202.5, 1203.3.7
F 1548—01(2006)	Standard Specification for the Performance of Fittings for Use with Gasketed Mechanical Couplings Used in Piping Applications	Table 1202.5
F 1807—13	Standard Specification for Metal Insert Fittings Utilizing a Copper Crimp Ring for SDR 9 Cross-linked Polyethylene (PEX) Tubing and SDR9 Polethylene of Raised Temperature (PE-RT) Tubing	Table 1202.5, Table 1210.5
F 1924—12	Standard Specification for Plastic Mechanical Fittings for Use on Outside Diameter Controlled Polyethylene Gas Distribution Pipe and Tubing	1210.6.6.3
F1960—12	Specification for Cold-Expansion Fittings with PEX Reinforcing Rings for Use with Cross-linked Polyethylene (PEX) Tubing	Table 1202.5
F 1974—09	Standard Specification for Metal Insert Fittings for Polyethylene/Aluminum/Polyethylene and Cross-linked Polyethylene/Aluminum/Cross-linked Polyethylene Composite Pressure Pipe	Table 1202.5
F2080—12	Specification for Cold-Expansion Fittings with Metal Compression-Sleeves for Cross-linked Polyethylene (PEX) Pipe	Table 1202.5
F2098—08	Standard Specification for Stainless Steel Clamps for SDR9 Cross-linked Polyethylene (PEX) Tubing to Metal and Plastic Insert Fittings	Table 1202.5
F 2159—11	Standard Specification for Plastic Insert Fittings Utilizing a Copper Crimp Ring for SDR 9 Cross-linked Polyethylene (PEX) Tubing and SDR9 Polyethylene of Raised Temperature (PE-RT) Tubing	Table 1210.5, Table 1202.5
F 2389—10	Specification for Pressure-rated Polypropylene Piping Systems	Table 1210.4, 1203.15.1, 1210.6.7.1, Table 1210.5, Table 1202.4, Table 1202.5

ASTM—continued

F2434—09	Standard Specification for Metal Insert Fittings Utilizing a Copper Crimp ring for SDR9 Cross-linked Polyethylene (PEX) Tubing and SDR9 Cross-linked Polyethylene/Aluminum/ Cross-linked Polyethylene (PEX-AL-PEX) Tubing	Table 1210.5
F 2623—08	Standard Specification for Polyethylene of Raised Temperature (PE-RT) SDR 9 Tubing1	Table 1202.4, Table 1210.4
F 2735—09	Standard Specification for Plastic Insert Fittings for SDR 9 Cross-linked Polyethylene (PEX) and Polyethylene of Raised Temperature (PE-RT) Tubing	Table 1202.5
F 2769—10	Polyethylene of Raised Temperature (PE-RT) Plastic Hot- and Cold-water Tubing and Distribution Systems	Table 1210.5, Table 1202.4
F 2806—10	Standard Specification for Acrylonitrile-Butadiene-Styrene (ABS) Plastic Pipe (Metric SDR-PR)	Table 1202.4

AWS

American Welding Society
8669 NW 36 Street, #130
Doral, FL 33166

Standard reference number	Title	Referenced in code section number
A5.8M/A5.8—2011	Specifications for Filler Metals for Brazing and Braze Welding	1203.3.1, 1303.3.1

AWWA

American Water Work Association
6666 West Quincy Avenue
Denver, CO 80235

Standard reference number	Title	Referenced in code section number
C901—08	Polyethylene (PE) Pressure Pipe and Tubing, $1/2$ in. (13mm) through 3 in. (76mm) for Water Service	Table 1210.4
C110/A21.10—12	Standard for Ductile Iron & Gray Iron Fittings	Table 1202.5
C115/A21.15—11	Standard for Flanged Ductile-Iron Pipe with Ductile Iron or Grey-Iron Threaded Flanges	Table 1202.4
C151/A21.51—09	Standard for Ductile-Iron Pipe, Centrifugally Cast for Water	Table 1202.4
C153/A21.53—11	Standard for Ductile-Iron Compact Fittings for Water Service	Table 1202.5

CSA

CSA Group
8501 East Pleasant Valley Road
Cleveland, OH 44131-5516

Standard reference number	Title	Referenced in code section number
B137.2—13	Polyvinylchloride (PVC) Injection-moulded Gasketed Fittings for Pressure Applications	Table 1210.5
B137.3—13	Rigid Poly (Vinyl Chloride) (PVC) Pipe for Pressure Applications	Table 1210.5
B137.6—13	Chlorinated Polyvinylchloride (CPVC) Pipe, Tubing and Fittings for Hot- and Cold-water Distribution Systems	Table 1210.5
B137.9—13	Polyethylene/Aluminum/Polyethylene (PE-AL-PE) Composite Pressure-Pipe Systems	Table 1210.4, Table 1210.5, Table 1202.4
B137.10—13	Cross-linked Polyethylene/Aluminum/Cross-linked Polyethylene (PEX-AL-PEX) Composite Pressure-Pipe Systems	Table 1202.4
CSA C448 Series—02 CAN/CSA 2002	Design and Installation of Earth Energy Systems—First Edition; Update 2: October 2009; Consolidated Reprint 10/2009	Table 1210.5
CSA B137.1—13	Polyethylene (PE) Pipe, Tubing and Fittings for Cold Water Pressure Services	Table 1210.4, Table 1210.5
CSA B137.5—13	Cross-linked Polyethylene (PEX) Tubing Systems for Pressure Applications	Table 1210.4, Table 1210.5
CSA B137.11—13	Polypropylene (PP-R) Pipe and Fittings for Pressure Applications	Table 1210.4, Table 1210.5
ANSI CSA America FC1—2012	Stationary Fuel Cell Power Systems	924.1

DOL

Department of Labor
Occupational Safety and Health Administration
c/o Superintendent of Documents
US Government Printing Office
Washington, DC 20402-9325

Standard reference number	Title	Referenced in code section number
29 CFR Part 1910.1000 (2009)	Air Contaminants	502.6
29 CFR Part 1910.1025 (2009)	Toxic and Hazardous Substances	502.19

FS

Federal Specifications*
General Services Administration
7th & D Streets
Specification Section, Room 6039
Washington, DC 20407

Standard reference number	Title	Referenced in code section number
WW-P-325B (1976)	Pipe, Bends, Traps, Caps and Plugs; Lead (for Industrial Pressure and Soil and Waste Applications)	Table 1202.4

*Standards are available from the Supt. of Documents, U.S. Government Printing Office, Washington, DC 20402-9325

ICC

International Code Council, Inc.
500 New Jersey Ave, NW
6th Floor
Washington, DC 20001

Standard reference number	Title	Referenced in code section number
IBC—15	International Building Code®	201.3, 202, 301.15, 301.16, 301.17, 301.18, 302.1, 302.2, 304.8, 304.11, 308.4.2.2, 308.4.2.4, 401.4, 401.5, 406.1, 501.3.1, 501.3.2, 501.10.2, 502.10, 502.10.1, 504.2, 504.10, 505.3, 506.3.3, 506.3.10, 506.3.12.2, 506.4.1, 509.1, 510.6, 510.6.3, 510.6.2, 510.7, 510.7.1.1, 510.7.2, 510.7.3, 510.8, 511.1.5, 513.1, 513.2, 513.3, 513.4.3, 513.5, 513.5.2, 513.5.2.1, 513.5.3, 513.5.3.2, 513.6.2, 513.10.5, 513.11.1, 513.12, 513.12.2, 513.20, 601.3, 602.2, 602.2.1.5.1, 602.2.1.5.2, 602.2.1.6.1, 602.2.1.6.2, 602.3, 602.4, 603.1, 603.10, 603.13, 603.18.2, 604.5.4, 607.1.1, 607.1.2, 607.3.2.1, 607.5.1, 607.5.2, 607.5.3, 607.5.4, 607.5.4.1, 607.5.5, 607.5.5.1, 607.5.6, 607.6, 607.6.1, 607.6.2, 607.6.2.1, 607.6.3, 701.2, 701.4.1, 701.4.2, 801.3, 801.16.1, 801.18.4, 801.18.4.1, 902.1, 908.3, 908.4, 910.3, 924.1, 925.1, 926.1, 927.2, 928.1, 1004.6, 1105.1, 1206.4, 1210.8.2, 1305.2.1, 1402.4, 1402.4.1
IECC—15	International Energy Conservation Code®	301.2, 303.3, 312.1, 401.2, 514.1, 604.1, 1204.1, 1204.2
IFC—15	International Fire Code®	201.3, 310.1, 311.1, 502.4, 502.5, 502.7.2, 502.8.1, 502.9.1, 502.9.5, 502.9.5.2, 502.9.5.3, 502.9.8.2, 502.9.8.3, 502.9.8.5, 502.9.8.6, 502.9.11, 502.10, 502.10.3, 502.16.2, 509.1, 510.2.1, 510.2.2, 510.5, 511.1.1, 513.1, 513.2, 513.6.3, 513.12.1, 513.12.3, 513.12.4, 513.15, 513.16, 513.17, 513.18, 513.19, 606.2.1, 606.4.1, 908.7, 924.1, 926.1, 1101.9, 1105.3, 1105.9, 1106.5, 1106.6, 1301.1, 1301.2, 1301.5
IFGC—15	International Fuel Gas Code®	101.2, 201.3, 301.6, 701.1, 801.1, 901.1, 906.1, 926.1, 1101.5
IPC—15	International Plumbing Code®	201.3, 301.11, 307.2.2, 512.2, 908.5, 928.1, 1002.1, 1002.2, 1002.3, 1005.2, 1006.6, 1008.2, 1009.3, 1101.4, 1201.1, 1206.2, 1206.3, 1210.8.1, 1401.2
IRC—15	International Residential Code®	101.2

IIAR

International Institute of Ammonia Refrigeration
1110 North Glebe Road
Arlington, VA 22201

Standard reference number	Title	Referenced in code section number
2—2014	Standard for Safe Design of Closed-circuit Ammonia Refrigeration Systems	1101.6, 1105.6.3

MSS

Manufacturers Standardization Society of the Valve & Fittings Industry, Inc.
127 Park Street, N.E.
Vienna, VA 22180

Standard reference number	Title	Referenced in code section number
SP 58—2009	Pipe Hangers and Supports—Materials Design and Manufacture, Selection, Application and Installation	305.4

NAIMA

North American Insulation Manufacturers Association
44 Canal Center Plaza, Suite 310
Alexandria, VA 22314

Standard reference number	Title	Referenced in code section number
AH116—09	Fibrous Glass Duct Construction Standards	603.5, 603.9

NBBI

National Board of Boiler and Pressure Vessel Inspectors
1055 Crupper Avenue
Columbus, OH 43229-1183

Standard reference number	Title	Referenced in code section number
NBIC—2011	National Board Inspection Code, Part 3	1003.3

NFPA

National Fire Protection Association
1 Batterymarch Park
Quincy, MA 02169-7471

Standard reference number	Title	Referenced in code section number
30A—15	Code for Motor Fuel-dispensing Facilities and Repair Garages	304.6
31—11	Standard for the Installation of Oil-burning Equipment	701.1, 801.2.1, 801.18.1, 801.18.2, 920.2, 922.1, 1308.1
37—15	Standard for the Installation and Use of Stationary Combustion Engines and Gas Turbines	915.1, 915.2
58—14	Liquefied Petroleum Gas Code	502.9.10
69—14	Standard on Explosion Prevention Systems	510.9.3
70—14	National Electrical Code	301.7, 306.3.1, 306.4.1, 511.1.1, 513.11, 513.12.2, 602.2.1.1, 927.2, 1104.2.2, 1106.3, 1106.4
72—13	National Fire Alarm and Signaling Code	606.3
82—14	Standard on Incinerators and Waste and Linen Handling Systems and Equipment	601.1
85—15	Boiler and Combustion Systems Hazards Code	1004.1
91—15	Standard for Exhaust Systems for Air Conveying of Vapors, Gases, Mists, and Noncombustible Particulate Solids	502.9.5.1, 502.17
92—15	Standard for Smoke Control Systems	513.7, 513.8
96—14	Standard for Ventilation Control and Fire Protection of Commercial Cooking Operations	507.1
211—13	Standard for Chimneys, Fireplaces, Vents and Solid Fuel-burning Appliances	806.1
262—15	Standard Method of Test for Flame Travel and Smoke of Wires and Cables for Use in Air-handling Spaces	602.2.1.1

NFPA—continued

286—15	Standard Methods of Fire Tests for Evaluating Contribution of Wall and Ceiling Interior Finish to Room Fire Growth	602.2.1.6.2
704—12	Standard System for Identification of the Hazards of Materials for Emergency Response	502.8.4, Table 1103.1, 510.1
853—15	Standard on Installation of Stationary Fuel Power Plants	924.1

NSF

NSF International
789 Dixboro Road, P. O. Box 130140
Ann Arbor, MI 48113-0140

Standard reference number	Title	Referenced in code section number
NSF 14—2011	Plastic Piping System Components and Related Materials	301.4
NSF 358-1—2011	Polyethylene Pipe and Fittings for Water Based Ground Source "Geothermal" Heat Pump Systems	Table 1210.4, Table 1210.5

SMACNA

Sheet Metal & Air-Conditioning Contractors National Assoc., Inc.
4201 Lafayette Center Drive
Chantilly, VA 20151-1209

Standard reference number	Title	Referenced in code section number
SMACNA/ANSI—2005	HVAC Duct Construction Standards-Metal and Flexible 3rd Edition (ANSI)	603.4, Table 603.4, 603.9, 603.10
SMACNA—10	Fibrous Glass Duct Construction Standards	603.5, 603.9

UL

UL LLC
333 Pfingsten Road
Northbrook, IL 60062-2096

Standard reference number	Title	Referenced in code section number
17—2008	Vent or Chimney Connector Dampers for Oil-fired Appliances— with revisions Through January 2010	803.6
103—2010	Factory-built Chimneys, Residential Type and Building Heating Appliance— with revisions through July 2012	805.2
127—2011	Factory-built Fireplaces	903.1, 903.3, 903.4
174—04	Household Electric Storage Tank Water Heaters—with revisions through September 2012	1002.1
180—2012	Liquid-level Indicating Gauges for Oil Burner Fuels and Other Combustible Liquids	1306.4
181—05	Factory-made Air Ducts and Air Connectors— with revisions through October 2008	512.2, 603.5, 603.6.1, 603.6.2, 603.9, 604.13
181A—2013	Closure Systems for Use with Rigid Air Ducts and Air Connectors	603.9
181B—2013	Closure Systems for Use with Flexible Air Ducts and Air Connectors	603.9
197—10	Commercial Electric Cooking Appliances—with revisions through June 2011	917.1
207—2009	Refrigerant-containing Components and Accessories, Nonelectrical	1101.2
263—2011	Standard for Fire Test of Building Construction and Materials	607.5.2, 607.5.5, 607.6.1, 607.6.2.1
268—2009	Smoke Detectors for Fire Alarm Systems	606.1
268A—2008	Smoke Detectors for Duct Application—with revisions through September 2009	606.1
343—2008	Pumps for Oil-Burning Appliances—with revisions through June 2013	1302.7
378—06	Draft Equipment—with revisions through January 2010	804.3, 804.3.8
391—2010	Solid-fuel and Combination-fuel Central and Supplementary Furnaces— with revisions through March 2010	918.1
412—2011	Refrigeration Unit Coolers—with revisions through August 2012	1101.2
471—2010	Commercial Refrigerators and Freezers—with revisions through December 2012	1101.2
499—05	Electric Heating Appliances—with revisions through February 2013	912.1, 923.1
508—99	Industrial Control Equipment—with revisions through March 2013	307.2.3
536—97	Flexible Metallic Hose—with revisions through June 2003	1302.8

UL—continued

555—06	Fire Dampers—with revisions through May 2012	607.3.1
555C—06	Ceiling Dampers—with revisions through May 2010	607.3.1
555S—99	Smoke Dampers—with revisions through May 2012	607.3.1
586—2009	High-Efficiency, Particulate, Air Filter Units	605.2
641—2010	Type L Low-temperature Venting Systems—with revisions through May 2013	802.1
705—2004	Standard for Power Ventilators—with revisions through March 2012	504.5
710—2012	Exhaust Hoods for Commercial Cooking Equipment	507.1
710B—2011	Recirculating Systems	507.1, 507.2
723—2008	Standard for Test for Surface Burning Characteristics of Building Materials	510.9, 602.2, 602.2.1, 602.2.1.6, 602.2.1.6.2, 602.2.1.6.3, 602.2.1.7, 604.3, 1204.1
726—95	Oil-fired Boiler Assemblies—with revisions through April 2011	916.1, 1004.1
727—06	Oil-fired Central Furnace—with revisions through April 2010	918.1
729—03	Oil-fired Floor Furnaces—with revisions through August 2012	910.1
730—03	Oil-fired Wall Furnaces—with revisions through August 2012	909.1
731—95	Oil-fired Unit Heaters—with revisions through August 2012	920.1
732—95	Oil-fired Storage Tank Water Heaters—with revisions through April 2010	1002.1
737—2011	Fireplace Stoves	905.1
762—2010	Outline of Investigation for Power Ventilators for Restaurant Exhaust Appliances	506.5.1
791—06	Residential Incinerators—with revisions through April 2010	907.1
834—04	Heating, Water Supply and Power Boilers Electric—with revisions through January 2013	1004.1
842—07	Valves for Flammable Fluids—with revisions through October 2012	1307.1
858—05	Household Electric Ranges—with revisions through April 2012	917.1
867—2011	Electrostatic Air Cleaners—with revisions through February 2013	605.2
875—09	Electric Dry Bath Heater—with revisions through November 2011	914.2
896—93	Oil-burning Stoves—with revisions through August 2012	917.1, 922.1
900—04	Air Filter Units—with revisions through February 2012	605.2
907—94	Fireplace Accessories—with revisions through April 2010	902.2
923—2013	Microwave Cooking Appliances	917.1
959—2010	Medium Heat Appliance Factory-built Chimneys	805.5
1046—2010	Grease Filters for Exhaust Ducts—with revisions through January 2012	507.2.8
1240—2012	Electric Commercial Clothes—Drying Equipment—with revisions through October 2012	913.1
1261—01	Electric Water Heaters for Pools and Tubs—with revisions through July 2012	916.1
1453—04	Electric Booster and Commercial Storage Tank Water Heaters—with revisions through July 2011	1002.1
1479—03	Fire Tests of Through-penetration Firestops—with revisions through October 2012	506.3.11.2, 506.3.11.3
1482—2011	Solid-fuel Type Room Heaters	905.1
1618—09	Wall Protectors, Floor Protectors and Hearth Extensions—with revisions through May 2013	308.4.1, 903.2, 905.3
1777—2007	Chimney Liners—with revisions through July 2009	801.16.1, 801.18.4
1812—2013	Standard for Ducted Heat Recovery Ventilators	514.1
1815—2012	Standard for Nonducted Heat Recovery	514.1
1820—04	Fire Test of Pneumatic Tubing for Flame and Smoke Characteristics—with revisions through May 2013	602.2.1.3
1887—04	Fire Tests of Plastic Sprinkler Pipe for Visible Flame and Smoke Characteristics—with revisions through May 2013	602.2.1.2
1978—2010	Grease Ducts	506.3.2, 506.3.6
1995—2011	Heating and Cooling Equipment	908.1, 911.1, 918.1, 918.2, 1101.2
1996—2009	Electric Duct Heaters—with revisions through November 2011	911.1
2024—2011	Standard for Safety Optical-Fiber and Communications Cable Raceway—with revisions through April 2011	602.2.1.1
2043—2008	Fire Test for Heat and Visible Smoke Release for Discrete Products and their Accessories Installed in Air-handling Spaces	602.2.1.4.2
2158—97	Electric Clothes Dryers—with revisions through March 2009	913.1
2158A—2010	Outline of Investigation for Clothes Dryer Transition Duct	504.8.3
2162—01	Outline of Investigation for Commercial Wood-fired Baking Ovens-Refractory Type	917.1
2200—2012	Stationery Engine Generator Assemblies—with revisions through June 2013	915.1
2221—2010	Tests of Fire Resistive Grease Duct Enclosure Assemblies	506.3.11.3
2518—05	Air Dispersion System Materials	603.17
2523—09	Solid Fuel-fired Hydronic Heating Appliances—with revisions through February 2013	1002.1, 1004.1

Appendix A
Chimney Connector Pass-Throughs

(This appendix is informative and is not part of the code.)

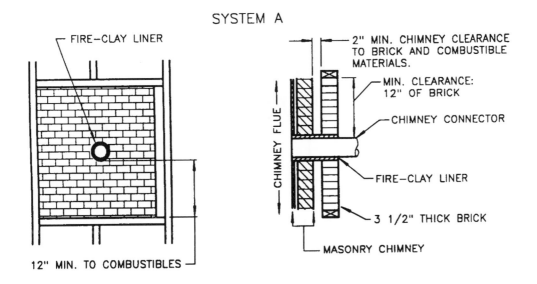

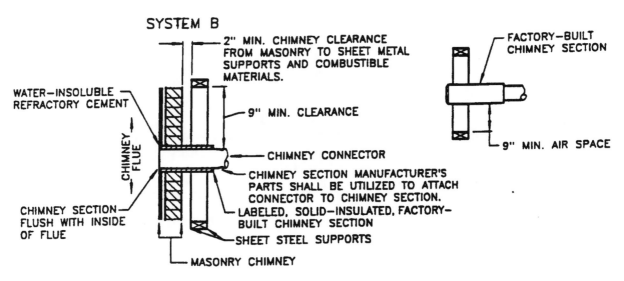

For SI: 1 inch = 25.4 mm.

FIGURE A-1
CHIMNEY CONNECTOR SYSTEMS

(continued)

SYSTEM C

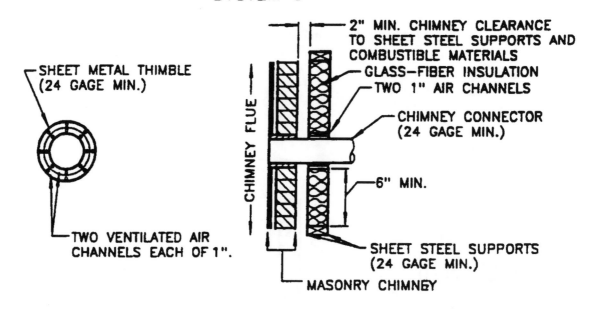

SHEET METAL THIMBLE (24 GAGE MIN.)

TWO VENTILATED AIR CHANNELS EACH OF 1".

CHIMNEY FLUE

2" MIN. CHIMNEY CLEARANCE TO SHEET STEEL SUPPORTS AND COMBUSTIBLE MATERIALS

GLASS-FIBER INSULATION

TWO 1" AIR CHANNELS

CHIMNEY CONNECTOR (24 GAGE MIN.)

6" MIN.

SHEET STEEL SUPPORTS (24 GAGE MIN.)

MASONRY CHIMNEY

SYSTEM D

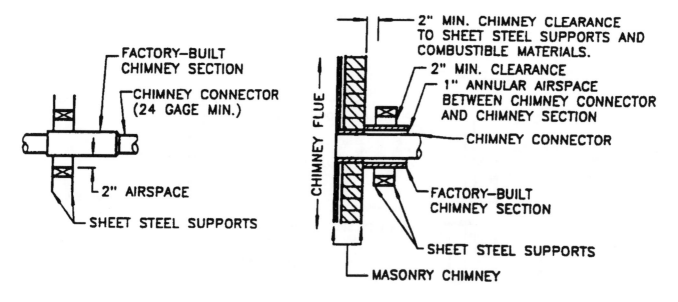

FACTORY-BUILT CHIMNEY SECTION

CHIMNEY CONNECTOR (24 GAGE MIN.)

2" AIRSPACE

SHEET STEEL SUPPORTS

CHIMNEY FLUE

2" MIN. CHIMNEY CLEARANCE TO SHEET STEEL SUPPORTS AND COMBUSTIBLE MATERIALS.

2" MIN. CLEARANCE

1" ANNULAR AIRSPACE BETWEEN CHIMNEY CONNECTOR AND CHIMNEY SECTION

CHIMNEY CONNECTOR

FACTORY-BUILT CHIMNEY SECTION

SHEET STEEL SUPPORTS

MASONRY CHIMNEY

For SI: 1 inch = 25.4 mm.

FIGURE A-1—continued
CHIMNEY CONNECTOR SYSTEMS

Appendix B
Recommended Permit Fee Schedule

(This appendix is informative and is not part of the code.)

B101
MECHANICAL WORK, OTHER THAN GAS PIPING SYSTEMS

B101.1 Initial Fee

For issuing each permit $___

B101.2 Additional Fees

B101.2.1 Fee for inspecting heating, ventilating, ductwork, air-conditioning, exhaust, venting, *combustion* air, pressure vessel, solar, fuel oil and refrigeration systems and *appliance* installations shall be $___ for the first $1,000.00, or fraction thereof, of valuation of the installation plus $___ for each additional $1,000.00 or fraction thereof.

B101.2.2 Fee for inspecting repairs, alterations and additions to an existing system shall be $___ plus $___ for each $1,000.00 or fraction thereof.

B101.2.3 Fee for inspecting boilers (based upon Btu input):

33,000 Btu (1 bhp) to 165,000 (5 bhp)	$ ___
165,001 Btu (5 bhp) to 330,000 (10 bhp)	$ ___
330,001 Btu (10 bhp) to 1,165,000 (52 bhp)	$ ___
1,165,001 Btu (52 bhp) to 3,300,000 (98 bhp)	$ ___
Over 3,300,000 Btu (98 bhp)	$ ___

For SI:1 British thermal unit = 0.2931 W, 1 bhp = 33,475 Btu/hr

B102
FEE FOR REINSPECTION

If it becomes necessary to make a reinspection of a heating, ventilation, air-conditioning or refrigeration system, or boiler installation, the installer of such *equipment* shall pay a reinspection fee of $___.

B103
TEMPORARY OPERATION INSPECTION FEE

When preliminary inspection is requested for purposes of permitting temporary operation of a heating, ventilating, refrigeration, or air-conditioning system, or portion thereof, a fee of $___ shall be paid by the contractor requesting such preliminary inspection. If the system is not *approved* for temporary operation on the first preliminary inspection, the usual reinspection fee shall be charged for each subsequent preliminary inspection for such purpose.

B104
SELF-CONTAINED UNITS LESS THAN 2 TONS

In all buildings, except one- and two-family dwellings, where self-contained air-conditioning units of less than 2 tons (7.034 kW) are to be installed, the fee charged shall be that for the total cost of all units combined (see B101.2.1 for rate).

INDEX

W

EDITORIAL CHANGES – SECOND PRINTING

Page 22, Section [BS] 301.16.1 now reads . . . **[BS] 301.16.1 Coastal high-hazard areas and coastal A zones.** In coastal high-hazard areas and coastal A zones, mechanical systems and *equipment* shall not be mounted on or penetrate walls intended to break away under flood loads.

Page 25, TABLE 305.4: column 1, row 15 now reads . . . PE-RT ≤ 1 inches

Page 35, Section 404.3: line 5 now reads . . . accordance with Section 403.3.1.

Page 37, TABLE 403.3.1.1—continued: column 2, row 2 now reads . . .

OCCUPANT DENSITY #/1000 FT² ª
—
10
50
20
120
30
120

Page 42, Section [F] 502.4: exception now reads . . . **Exception:** Lithium-ion and lithium metal polymer batteries shall not require additional ventilation beyond that which would normally be required for human occupancy of the space.

Page 42, Section 502.4.3: section number now reads . . . [F] 502.4.3.

Page 42, Section 502.5.3: section number now reads . . . [F] 502.5.3.

Page 63, Section [F] 513.5: 1st paragraph now reads . . . **[F] 513.5 Smoke barrier construction.** Smoke barriers required for passive smoke control and a smoke control system using the pressurization method shall comply with Section 709 of the *International Building Code*. The maximum allowable leakage area shall be the aggregate area calculated using the following leakage area ratios:

Page 64, Section 513.5.3.1 section number now reads . . . [F] 513.5.3.1.

Page 64, Section [F] 513.6.2: Equation 5-2 now reads . . . (Equation 5-1).

Page 65, Section [F] 513.10.1: Equation 5-4 now reads . . . (Equation 5-2).

Page 66, Section [F] 513.11.1 now reads . . . **[F] 513.11.1 Equipment room.** The standby power source and its transfer switches shall be in a room separate from the normal power transformers and switch gear and ventilated directly to and from the exterior. The room shall be enclosed with not less than 1-hour fire-resistance-rated fire barriers constructed in accordance with Section 707 of the *International Building Code* or horizontal assemblies constructed in accordance with Section 711 of the *International Building Code*, or both.

Page 77, Section [BF] 607.5.2: line 7 now reads . . . 1024.6, respectively, of the *International Building Code*.

Page 88, new section added now reads . . . **805.7 Factory-built fireplaces.** Chimneys for use with factory-built fireplaces shall comply with the requirements of UL 127.

Page 109, Section [F] 1109.1: Item 3, line 2 now reads . . . emergency ventilation systems.

Page 119, TABLE 1302.3: column 2, row 3, line 2 now reads . . . ASTM B 280; ASME B 16.51

Page 120, Section 1303.7 now reads . . . **1303.7 Copper or copper-alloy tubing.** Joints between copper or copper-alloy tubing or fittings shall be brazed, mechanical joints complying with Section 1303.3, press connect joints that conform to one of the standards in Table 1302.3 or flared joints. Flared joints shall be made by a tool designed for that operation.

Page 131, reference standard IIAR title now reads . . . Standard for Safe Design of Closed-circuit Ammonia Refrigeration Systems

Page 132, reference standard NSF: new standard added now reads . . . NSF 14—2011 Plastic Piping System Components and Related Materials. . .301.4

Page 132, reference number SMACNA/ANSI—2015 now reads . . . SMACNA/ANSI—2005

EDITORIAL CHANGES – THIRD PRINTING

Page 25, TABLE 305.4: column 1, row 15 now reads . . . PE-RT 1 inch and smaller

Page 25, TABLE 305.4: column 1, row 16 now reads . . . PE-RT $1^1/_4$ inches and larger

Page 25, TABLE 305.4: column 1, row 18 now reads . . . Polypropylene (PP) pipe or tubing, 1 inch and smaller

Page 25, TABLE 305.4: column 1, row 19 now reads . . . Polypropylene (PP) pipe or tubing, $1^1/_4$ inches and larger

Page 34, Section 403.3.2.1: **Equation 4-9** now reads . . . $Q_{OA} = 0.01 A_{floor} + 7.5(N_{br} + 1)$ **(Equation 4-9)**

Page 75, Section 607.1.1 title now reads . . . **[BF] 607.1.1 Ducts and air transfer openings.**

Page 131, NFPA: referenced standard 37—14 now reads . . . 37-15.

Page 132, SMACNA/ANSI—2005 title now reads . . . HVAC Duct Construction Standards-Metal and Flexible 3rd Edition (ANSI)

INTERNATIONAL CODE COUNCIL®

People Helping People Build a Safer World®

Valuable Guides to Changes in the 2015 I-Codes®

NEW!

FULL COLOR! HUNDREDS OF PHOTOS AND ILLUSTRATIONS!

SIGNIFICANT CHANGES TO THE 2015 INTERNATIONAL CODES®

Practical resources that offer a comprehensive analysis of the critical changes made between the 2012 and 2015 editions of the codes. Authored by ICC code experts, these useful tools are "must-have" guides to the many important changes in the 2015 International Codes.

SIGNIFICANT CHANGES TO THE IBC, 2015 EDITION
#7024S15

SIGNIFICANT CHANGES TO THE IRC, 2015 EDITION
#7101S15

SIGNIFICANT CHANGES TO THE IFC, 2015 EDITION
#7404S15

SIGNIFICANT CHANGES TO THE IPC/IMC/IFGC, 2015 EDITION
#7202S15

Changes are identified then followed by in-depth discussion of how the change affects real world application. Photos, tables and illustrations further clarify application.

ORDER YOUR HELPFUL GUIDES TODAY!
1-800-786-4452 | www.iccsafe.org/books

HIRE ICC TO TEACH

Want your group to learn the Significant Changes to the I-Codes from an ICC expert instructor? Schedule a seminar today!
email: **ICCTraining@iccsafe.org** | phone: **1-888-422-7233 ext. 33818**

14-09379

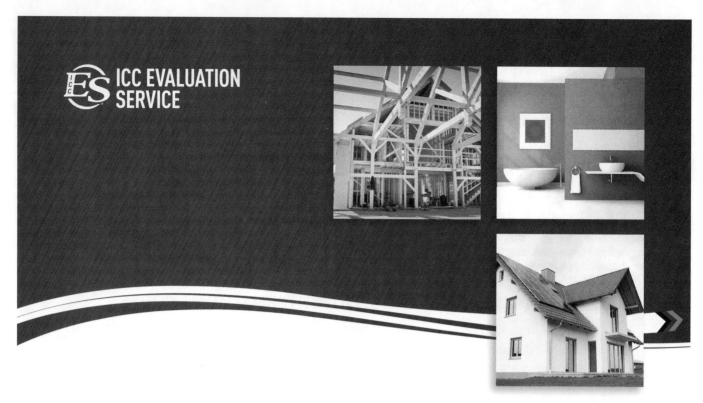

APPROVE WITH
CONFIDENCE

- ICC-ES® Evaluation Reports are the most widely accepted and trusted in the nation.
- ICC-ES is dedicated to providing evaluation reports, PMG and Building Product Listings to the highest quality and technical excellence.
- ICC-ES is a subsidiary of ICC®, the publisher of the IBC®, IRC®, IPC® and other I-Codes®.

We do thorough evaluations. You approve with confidence.

Look for the ICC-ES marks of conformity before approving for installation

 Code Officials — Contact Us With Your Questions About ICC-ES Reports at:
800-423-6587 ext. ES help (374357) or email EShelp@icc-es.org
www.icc-es.org

14-09329

Subsidiary of

People Helping People Build a Safer World®

The International Code Council's Training and Education Department

Industry professionals look to ICC to provide the critical knowledge and experiences necessary to excel in today's challenging world. ICC's educational programs cover a broad spectrum of code and construction related topics, offering some of the highest quality instruction direct from the source.

INSTITUTES

Acquire Skills, Share Experiences, Earn CEUs

ICC's training institutes offer a comprehensive education experience and a great way to earn much needed CEUs. Learn best practices from the leading experts in the field, build your network of professional contacts and share experiences. Institutes are held across the country throughout the year. To see the full schedule go to www.iccsafe.org/training.

HIRE ICC TO TEACH

Bring ICC On-Site and Earn CEUs

Give your group the confidence they need to meet tough challenges head-on so they can reach their full potential. ICC's course catalog contains a wide variety of educational topics that are available for contract at a location of your choice. Customized training is also available to fit your needs. For our full course catalogue go to www.iccsafe.org/hireicc.

ICC ONLINE CAMPUS

Earn CEUs at Your Own Pace

Online courses offer access to a comprehensive training and education portal, providing you with an effective and convenient tool to enhance your professional skills. Courses are available anytime for 99 days following registration. For a quick and easy way to earn CEUs towards your certification, try our Online Certifications Renewal section. Go to www.iccsafe.org/onlinecampus for our full offerings.

For more information about ICC's Training please contact us at
888-ICC-SAFE (422-7233) ext. 33818 or email us at **icctraining@iccsafe.org.**

14-09372

INTERNATIONAL CODE COUNCIL®

People Helping People Build a Safer World®

Growing your career is what ICC Membership is all about

As your building career grows, so does the need to expand your code knowledge and job skills. Whether you're seeking a higher level of certification or professional quality training, Membership in ICC offers the best in I-Code resources and training for growing your building career today and for the future.

- **Learn** new job skills to prepare for a higher level of responsibility within your organization
- **Improve** your code knowledge to keep pace with the latest International Codes (I-Codes)
- **Achieve** additional ICC Certifications to open the door to better job opportunities

Plus, an affordable ICC Membership provides exclusive Member-only benefits including:

- Free code opinions from I-Code experts
- Free I-Code book(s) to new Members*
- Access to employment opportunities in the ICC Career Center
- Discounts on professional training & Certification renewal exams
- Savings of up to 25% off on code books & training materials
- Free benefits - Governmental Members: Your staff can receive free ICC benefits too*
- And much more!

Join the International Code Council (ICC) and start growing your building career now! Visit our Member page at **www.iccsafe.org/membership** for an application.

*Some restrictions apply. Speak with an ICC Member Services Representative for details.

14-09333

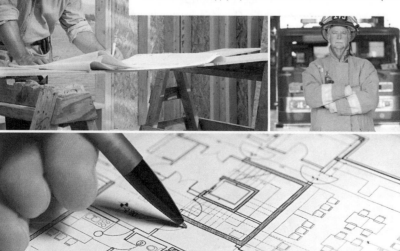